ORIGINS OF MENDELISM

Origins of Mendelism

ROBERT C. OLBY

With an Introduction by
Professor C. D. Darlington

SCHOCKEN BOOKS · NEW YORK

Library of Congress Catalog Card No. 66-20463

Second Printing, 1967

Manufactured in the United States of America

To

DR. JAROSLAV KŘÍŽENECKÝ

(1896 — 1964)

First Director

of the Gregor Mendel

Department of Genetics in the

Moravian Museum, Brno, Czechoslovakia

Foreword

by Professor C. D. Darlington

Each generation that looks at Mendel's experiments will, it has been said, find something new in them. This indeed seems to be true. But now, a hundred years after his paper appeared, we can go further. We can ask about the people like Gaertner and Thomas Andrew Knight and Charles Naudin. They came before Mendel and made the same kind of experiments that he made without understanding them. We can also ask about the people like Carl Naegeli and Karl Pearson who came after Mendel and saw his results but misunderstood and rejected them. We can even ask about those like Darwin and Galton who examined Mendel's problem in the twilight years between his work and its rediscovery. They understood the possibilities but did not know how to make the experiments which would test them.

We can take a glance, too, at the writings of Weismann, who independently of Mendel and in these same twilight years invented his own theory of heredity. It was atomic heredity and hard heredity, as hard as Mendelism itself. But for Weismann it was a theory of the cell to be studied only by the microscope. And he did not dream that it could be vindicated and established by breeding experiments.

Such vicissitudes of observation, experiment and theory have never occurred at the birth of any other science. But the strange history of Mendelism has made it possible for Robert Olby in this book to consider them all. In doing so he has brought to light many new things. Some bear on the politics of publication and the mystery of how three investigators could simultaneously rediscover Mendel. Others bear on the general theory of heredity. They help to clear away misunderstandings which are still as widespread as ever. They also bear on the processes of discovery. Who would have imagined for example that Galton could have set out a Mendelian ratio in a letter to Darwin? Or that when the ratio had been explained to him the whole matter should be overlooked by Darwin and by Galton himself. And that it should only be recalled by Galton's biographer Karl Pearson fifty years later?

Such a revelation goes deeper than the mere history of biology. It tells us something new about what may happen in the minds of discoverers. Or fail to happen. Readers in many fields of science will, I believe, be indebted to Dr. Olby for having made these inquiries and for the simplicity with which he has described their results.

C.D.D.

Author's Note

In 1929 H. F. Roberts published an excellent book on the work of the pre-Mendelian hybridists. There he pointed out the fact that genetic dominance and hybrid segregation were well known long before Mendel began his experiments. Unfortunately his book is little known today, so it is perhaps not surprising that Mendel's work is often regarded as an isolated achievement, for the success of which he owed nothing to his predecessors or contemporaries. Thus a recent B.B.C. Third Programme talk began with the statement that:

> The science of genetics is exactly 100 years old. The fact that it is possible to make so blunt an assertion illustrates an important principle in the history of science. Often, as Newton said, the successor can see further than his precursors because, standing on their shoulders, he can see a little further than they. There was nothing of this kind about Gregor Mendel. He had no precursors to stand on at all. It is true that before him there were breeders of plants like Sageret and Naudin, and breeders of mice like Colladon, who made observations which can now, with the help of hind-sight, be seen as parts of the Mendelian principle. But none of them succeeded in finding that principle: Mendel had no knowledge of their work, and his work would have been in no way different if they had never existed at all.
>
> (*Listener*, *73*, No. 1876, 364. March 11, 1965)

I see the work of Mendel as the culmination of a series of studies which began two centuries ago with the classic hybridisation experiments of Koelreuter. In this book I have sought to establish this point by going over much of the ground already studied by H. F. Roberts and by adding to his work. I have also attempted to look at the subject in its wider aspects in order to describe the growth of ideas about inheritance and variation which eventually led to the Mendelian solution. In this sense the book may be said to be about the origins of Mendelism.

Robert C. Olby

The Botany School, Oxford.
April 1965.

Contents

List of Plates

The following plates are reproduced by courtesy of:

1. Badisches Generallandesarchiv, Karlsruhe.
3a. Walter Staudenmeyer. Stadtarchivist, Kreisstadt Calw.
3b. Verlag Kurt Ziegler, Calw.
6. Cambridge University Library.
7, 8, 9, 10, 11, 12. Gregor Mendel Department of Genetics, Moravian Museum, Brno.

The author is grateful to Walter Staudenmeyer for the note on Achatius Gaertner in the caption to Plate 3b.

The two passages in the appendices from Mendel's *Experiments in Plant Hybridisation* are reprinted by kind permission of the Council of the Royal Horticultural Society.

List of Tables

List of Abbreviations used in the Notes to the Chapters

Abbreviation	*Author*	*Place of full citation*	
		Chapter	*Note*
Au	Mendel	v	3
Bz	Gaertner	ii	9
HG	Galton	iii	31
Il	Iltis	v	5
LFG	Pearson	iii	27
LLD	Darwin	iv	27
MLD	Darwin	ii	2
MR1	Mendel	i	17
MR2	Mendel	i	18
MR3	de Vries	vi	13
N1–4	Darwin	iii	5
O1e	Darwin	iii	8
O6e	Darwin	iii	6
OK	Koelreuter	i	9
Roberts 1929	Roberts	vi	6
VuD	Darwin	iii	11

ORIGINS OF MENDELISM

Chapter One

The Early Hybridists

Ever since man first tried to penetrate beneath the outward appearances of nature in his search for patterns and regularities he has been trying to formulate a law which governs the resemblance of children to their parents. The Greek scientists made a number of suggestions. Some said it was a question of which sex dominated in the sexual act. Others held that it was simply a question of the heat of the womb or of which testis the sperm came from. Having thus disposed of the question of sex they went on to assert that in other characteristics boys are like their fathers and girls are like their mothers.

In this way the question of the inheritance of those characteristics which have nothing to do with sex were confused with it, and this was so not only in the writings of Aristotle but also in those of eighteenth- and nineteenth-century authors. Thus the great French naturalist, George Louis Leclerc de Buffon (1707–1788), held that the male determines the extremities of the body—head, tail and limbs—and the female determines the internal parts and the overall shape and size. In support of this assertion he described in detail the characteristics of the offspring from seven different crosses: ass × horse, mare × ass, wolf × mastiff, canary × goldfinch, siskin and linnet, ewe × goat.* With the exception of the last example Buffon obtained accounts of these hybrids from correspondents and in no case did he actually supervise

* In each case the female is cited first.

the hybridisation. Nevertheless his appeal to supposedly well-authenticated experiments should be noted for it is evidence of a general trend towards an insistence on experimental backing for statements about nature.

Buffon realised that he needed more evidence, and to this end he appealed to his readers to carry out experiments in hybridisation. But animal hybrids of the sort he envisaged are not easy to produce and the experiments require time, patience and money. Buffon and his contemporaries made life difficult for themselves by using wolves, horses and goats instead of such quick-breeding, prolific animals as rats and mice. It was unfortunate if the wolf killed the dog with whom she was intended to mate, or mauled the coachman so badly that she had to be killed.[1] But what was one to do? One could not use rats and mice—that would be breeding vermin! Thus did good taste restrict the choice of experimental material.

What about plants? Though they had for centuries been regarded as lowly forms of life which lack sexuality, their possession of this characteristic had been demonstrated convincingly in 1694 by Rudolph Camerarius, Professor of Natural History at Tübingen. Plants are many thousand times more prolific than dogs, goats and horses. They give rise to no emotional problems and they are far cheaper to feed. Also their pollination can be controlled with ease. Yet despite all these advantages no one hybridised them for strictly scientific purposes before 1750.[2] Of course the scientific world had first to accustom itself to the notion that plants have sex. It disturbed their tidy conception of the organic world according to which the insensitive and asexual plants form the base of a "chain of being" which ascends by imperceptible degrees through the various departments of the animal kingdom to the perfection of man. And pious naturalists, who expected to learn good morals from nature, were shocked that there should be such an abundance of pollen grains and so few "seed chambers", for this meant that there are always many more males than there are females. "What man", exclaimed J. G. Siegesbeck, Professor of Botany at St. Petersburg, "will ever believe that God Almighty should have introduced such confusion, or rather such shameful whoredom, for the propagation of the reign of plants. Who will instruct young students in such a voluptuous system without scandal?"[3]

This was just what the Swedish botanist, Carl Linnaeus (1707–1778) was doing at Uppsala. Not only did he uphold the new doctrine but he also made the sexual organs the basis of his system of plant classification. Such was the success of this system that its author became a national hero and the most renowned botanist of his century. Linnaeus, too, knew full well how successful his work had been and though he thanked God for his good fortune there is no mistaking the arrogance of the following autobiographical notes:

> God has permitted him (Linnaeus) to see more of his created work than any mortal before him.
>
> God has bestowed on him the greatest insight into nature-study . . .
>
> None before him has written more works, more correctly, more methodically, from his own experience.
>
> None before him has so totally reformed a whole science and made a new epoch. . . .
>
> None before him has sent out his disciples to so many parts of the world.[4]

Thus as the apostle of plant sexuality he sent his disciples into the fields to search for hybrids. If any plant showed characters which were intermediate between those of two known species, that was good enough for him: it must be a hybrid. There was no suggestion that it should be tested for purity of type. He assumed it bred true—that it was a true-breeding hybrid—a new species!

With this loose idea of hybrids it is not surprising that Linnaeus and his students imagined they had found a great number of them. And in a thesis of Linnaeus which his student Haartman presented in 1751 no less than 100 are listed.[5] But of these only about half a dozen could possibly have been hybrids, the remaining 94 being quite impossible combinations. Thus did Linnaeus forge ahead without stopping to consider whether carefully designed experiments would verify or falsify his conclusions.

In 1759 the Academy of Sciences at St. Petersburg offered a prize of 50 ducats, ducats and pounds sterling being roughly equivalent at the time, for an essay confirming or refuting the sexes of plants by fresh arguments and experiments. Linnaeus at once wrote an essay and in the autumn of 1760 he was duly awarded the prize. Reading this essay today one is impressed with Linnaeus' flowing style and his boldness in making assertions. He seems to have all the answers. Even the difficult question of the roles of male and female in reproduction is neatly answered by analogy with animals. The evidence from animal hybrids, he believed, pointed to a two-layer theory of heredity—the outer layer including the vascular system is derived from the father, the inner layer including the nervous system comes from the mother. In plants he held that the leaves and the rind of the stem constitute the paternal outer skin, whilst the central part of the flower, the "fructification", and the pith of the stem constitute the maternal core. Of the plant hybrids which he cited in support of this theory only two were described as being produced artificially by hand cross-pollination. They were: a hybrid goat's beard (*Tragopogon pratensis* × *T. porrifolius*) and a hybrid speedwell (*Veronica maritima* × *Verbena officinalis*), but the latter combination seems an extremely unlikely one.

It was certainly fortunate for Linnaeus that a botanist, working in St. Petersburg at the time trying to produce plant hybrids, did not succeed until after the award of the prize. This was the German Joseph Gottlieb Koelreuter (1733–1806). He raised his first hybrid plants in the autumn of 1760. They were tobaccos, and they came into flower in the following spring. In the autumn of 1761 he published a description of them in a little book called *Vorläufige Nachricht von einigen das Geschlecht der Pflanzen betreffenden Versuchen und Beobachtungen* (Preliminary report of experiments and observations concerning some aspects of the sexuality of plants). As the results of further experiments became available he reported these in three *Fortsetzungen* (Continuations) in 1763, 1764 and 1766.

This book never achieved a wide circulation but it is nevertheless a classic in the history of biology, and the experimental study of genetics may be said to date from the work which Koelreuter described in it. The text is chiefly devoted to a sober account of careful experiments in hybridisation, 65 in all, and of parallel investigations into the mechanisms of pollination and fertilisation. When we take into account the later work which he described in the journals of the St. Petersburg Academy of Sciences the total number of experiments, judged by eighteenth-century standards, is fantastic. Thus he carried out more than 500 different hybridisations involving 138 species, and examined the shape, colour and size of the pollen grains from over 1,000 different plant species.

Sad to relate, the records of all these experiments were passed over by almost all his contemporaries and forgotten. He did win a measure of local fame as a *savant* of Karlsruhe, where he was Professor of Natural History. A friend in St. Petersburg named a beautiful genus of trees *Koelreuteria* in his honour, and Hedwig named a genus of mosses after him. But Koelreuter himself was frustrated and bitter to the end of his days.

When Count Leopold von Stolberg visited him in 1791 he was most impressed by the Professor and thought his transmutation of one species of plant into another particularly interesting. He left the following account of his visit:

> By repeated experiments [in transmutation], he has no less successfully reduced these varieties to their original form and genus. He has again conducted them through their different gradations, and again and again fully restored them to all their original powers, and properties: bringing back some of them to the male kind, and others to the female.
>
> Tiresius was struck blind, when he daringly endeavoured to unfold the secrets of Venus. May we not expect that another Nemesis shall pursue the man who, with wonderful wisdom and passionate ardour, has drawn aside the veil of nature?
>
> This bold and discreet observer, who watched the bees at their employment, and who, by placing glass tubes in the ambrosial cups of flowers (nectaries), robbed them of their sweets and brought forth honey, this remarkable man has not a foot of land that he can call his own. Not one of the great men of Germany has conferred on himself the honour, or the delight, of bestowing a garden on this sage: whose science is as pleasant as it is abundantly beneficial.[6]

But Count von Stolberg's plea fell on unreceptive ground, for the eighteenth century was the age of Linnean botany, the age of great voyages and mammoth collections. Every biologist of international standing had been on an expedition and had collected something. The stay-at-home experimentalist was just not in the picture. His work was regarded as "curious" and "ingenious" but rarely as "important". Thus Linnaeus, in his celebrated *Philosophia Botanica* described the work of the plant anatomists Malpighi, Grew and Hales as "not properly belonging to botany as a science".[7] In the nineteenth century, however, the subjects which these men tackled came to assume an important place in botany. The same was true for Koelreuter's work. Apart from a repetition of one of his hybridisations by the Swiss bryologist Johann Hedwig in 1798, his experiments were not repeated until half a century after the publication of his *Vorläufige Nachricht* and twenty years after his death. Augustin Sageret (1763–1861), the French agronomist, was the first, but he gave no details of his experiments. He was followed by Dr. Wiegmann of Braunschweig and by Carl von Gaertner of Calw (1772–1850). They all testified to the accuracy of Koelreuter's work, much to the surprise of those critics who still denied the truth of plant sexuality and questioned the import of Koelreuter's experiments.

It was to silence these critics that Gaertner repeated and extended Koelreuter's work and thereby laid the foundations for the work of Mendel. Hence, continuity of genetic thought can be traced through the writing of these three men. In addition, all three were concerned with the question of the origin of species, and Charles Darwin made a special study of Gaertner's and Koelreuter's publications. In his famous book *On the Origin of Species . . .*, he referred frequently to their work, and only then did the name of Koelreuter become well known. In the present century the greater part of his work in the hybridisation of tobaccos was repeated and extended by the American botanists, Thomas Goodspeed and Edward East, who were able to account for the non-Mendelian nature of many of Koelreuter's results.

Koelreuter's Life

Joseph Koelreuter was born in 1733 in the little town of Sulz on the Neckar river in the Black Forest. His father was the local apothecary and when Joseph was fifteen he sent him to the ancient University of Tübingen to study medicine. Apart from a brief stay at Strasbourg University he remained at Tübingen until 1755 when he received his degree and left for the Russian capital of St. Petersburg. He worked as a natural historian in the Academy of Sciences there for six years. His official work was chiefly concerned with the classification of fishes and corals but he also found time to study the structure of flowers and the mechanisms of their pollination and to attempt plant hybridisation. In 1761, after he had produced his first hybrid plants, he left St. Petersburg for home, but it was not until the winter of 1763 that he received another appointment, this time to Karlsruhe, only about fifty miles from Sulz. Karlsruhe was the seat of the Margrave of Baden, Karl Friedrich (1749–1811), whose wife Caroline was an enthusiastic botanist. She tried to persuade Linnaeus to come to Karlsruhe as Professor of Natural History and Director of the Margrave's gardens, but having failed, Koelreuter was appointed.

Before his move to Karlsruhe Koelreuter had had to grow his hybrids wherever he could, either in pots which he took with him on his journeys, or in the gardens of his friends and relations. Some of his hybrids were grown in the garden of the Gaertner family in Calw. At Karlsruhe, however, he had a whole botanic garden and a staff of gardeners at his disposal. Forthwith he set to work to continue and extend his experiments. He wrote abroad for rare and exotic plants and seeds, which he wished to hybridise, provided at his own expense a glasshouse and fuel to accommodate them, but he left to the head gardener, Saul, and the Court Gardener, Mueller, the task of managing the house and tending the plants. These two viewed Koelreuter's experiments with singular distaste, and by simply ignoring his instructions they succeeded in ruining most of the experiments. The hybrids growing in the open were soon choked with weeds and those in the glasshouse died for want of heat. Saul seems to have been the real villain of the piece; he opposed Koelreuter's ruling on every little item on the daily agenda of work. He wrote to the financial committee (*Rentkammerkollegium*) of the gardens complaining about Koelreuter, who then wrote complaining of Saul. Nothing would induce the parties to be reconciled and eventually, when Koelreuter's friend and protector, Caroline, died in 1783, he was dismissed from the position of Director of the Gardens and his place taken by Carl C. Gmelin (1762–1837).

Koelreuter remained Professor of Natural History and continued to live in Karlsruhe, but his disagreement with Saul had far-reaching consequences. Many of his later experiments were terminated prematurely; his study of sexuality in the lower plants suffered from a lack of facilities and good equipment, and he was never able to carry out his intention of hybridising finches to show that his conclusions regarding plants apply also to animals. In his later years the sense of frustration grew. We find it creeping into his writings, and Von Stolberg, as we have seen, was given an account of his grievances in 1791.

In 1775 Koelreuter married the daughter of a local judiciary. She bore him six children. Gottlieb, Karl and Wilhelm were all given a university education at considerable expense to their father who had frequently to borrow money. In Karlsruhe he borrowed a total of 800 gulden and mortgaged his house twice. The eldest son, Gottlieb, spent only two years at Tübingen University before he wrote to the Margrave asking to be allowed to practise as a doctor in Karlsruhe. He was duly examined by a committee of doctors who naturally refused his request. A year later, in 1801, he died. Shortly after this Koelreuter's wife died. Koelreuter continued with his research until 1805 when he developed chest trouble which led to his death in 1806.

Genetic Researches

The sad story of Koelreuter's life contrasts markedly with the excitement of his scientific discoveries. To appreciate the great pleasure which these discoveries gave him we must go back to the year 1759 when he started his hybridisation experiments and his study of fertilisation and pollination. We have seen that at that time there were but two well-known plant hybrids produced by artificial cross-pollination, Linnaeus' speedwell and goat's beard hybrids. The former was produced in the Uppsala Botanic Gardens in 1750. The latter arose spontaneously in 1750 and was produced artificially in 1758, also in the Uppsala Botanic Gardens. Linnaeus' account of these crosses is very brief and in neither case did he state how many plants he grew, how fertile they were and whether he raised a second generation from them. All we know about their fertility is that the hybrid speedwell was propagated by cuttings and the hybrid goat's beard produced seeds which he sent to St. Petersburg. Koelreuter, who was working there at the time, raised plants from Linnaeus' seeds, but they were not like the original hybrids; they were, presumably, F2 hybrids.

Linnaeus, however, was sure they were both true-breeding species.[8] He believed that the speedwell hybrid had already occurred naturally outside the Uppsala Botanic Garden and was the plant which earlier botanists had called *Veronica spuria*. He named the hybrid goat's beard *Tragopogon hybridum*, and even before he had produced the latter artificially he entered it together with *Veronica spuria* in his highly respectable *Species Plantarum* (1753).

If he was justified in doing this, here was a bombshell to established doctrine, for nature was supposed to preserve the same order and harmony as had reigned in the garden of Eden. But if man can create new species whenever he chooses simply by hybridising existing species there would be no end to the confusion. Such a state of affairs was unthinkable to the pious naturalists of the eighteenth century, and Koelreuter, who did not doubt that plant hybrids can be produced, was sure that nature has her own ways of preventing them from producing fresh species. The chief aim of his work was to find out these hidden measures. Accordingly he examined the fertility of his hybrids with great care. His first hybrid *Nicotiana rustica* × *N. paniculata* grew so well that he began to entertain doubts as to its hybrid nature. "The keenest eye", he said, "can discern no imperfection, from the embryo to the more or less complete formation of its flowers."[9] But when all the flowers fell off, not a single fruit formed, and instead of the expected 50,000 seeds per plant there was none. This struck Koelreuter as "one of the most wonderful of all events that have ever occurred upon the wide field of nature".[10] It was also one of his most comforting discoveries for it banished the evil spectre of self-perpetuating hybrids at least temporarily from his view. At the same time he was perplexed by this remarkable contrast between the fertility of pure species and the sterility of hybrids. "All human understanding taken together", he said, "may be too weak to solve it", so he did not propose to "break his head on it".[11]

Although he did not attempt to untie this "most complicated of all knots" by reasoning, he attacked it by further experiments and observations. Thus he examined the hybrid pollen grains under the microscope and found them "shrunken and, as it were, pulverised; they contained scarcely any fluid material, and were, in a word, mere empty husks".[12] Clearly it was sterile, and to see whether the ovary was also sterile he tried the effect of pollinating some of the hybrids with pollen from *N. rustica* and others with pollen from *N. paniculata*. These back crosses were successful and from them he raised a second generation of hybrids, what we call back-crossed hybrids and what Koelreuter called hybrids of the first ascending or descending degree (descent being to the maternal species and ascent to the paternal species). Later he even succeeded in raising true second generation hybrids (what we call F2 hybrids) from the self-pollination of these tobacco hybrids. Not only the ovary but also the pollen had a slight degree of fertility. And when he came to the hybridisation of pinks, carnations, sweet williams (*Dianthus chinensis* & etc., *D. caryophyllus*, *D. barbatus*) and varieties of the marvel of Peru (*Mirabilis jalapa*) he found a much higher degree of fertility and was able thus to grow fairly representative numbers of F2 and back-crossed hybrids.

The result was that for the first time in the history of biology reliable and accurate descriptions of hybrids and of their descendants were available, and when Koelreuter compared them he found a striking contrast. F1 hybrids for any given cross were all alike and in most of their characters they were intermediate between the two parental species. F2 and back-crossed hybrids were all different, even those derived from one and the same ovary, and they tended to be less like their parental hybrids and more like one or other of the originating species.

This contrast between the two generations remained an enigma until 1900 when Mendel's explanation was made generally known. Whereas Mendel explained the enigma on cytological and statistical grounds Koelreuter explained it on bases which may be described as theological and alchemical.

Explanation of results

Koelreuter looked upon the wonderful uniformity and almost exact intermediacy of F1 hybrids as evidences of Nature's perfection. The same cross repeated no matter how many times gave the same result.[13] What caused the breakdown of this ordered and precise result in the second generation? Surely, he reasoned, it must be man. Nature never intended that species should be crossed and to prevent it she had placed closely related forms far apart. Then came man mixing up nature's careful arrangement and cramming into the confines of his little gardens species which formerly were separated by thousands of miles. Under these conditions that which nature never intended took place by the cross-pollinating action of insects, wind or man. Plants were thus produced in which were combined two specific materials "not intended for each other by the wise Creator"[14]. The unnatural state of these unions became evident not in the vegetative faculty which was enhanced but in the reproductive faculty which was partially if not wholly impaired. In cases where this reduction of fertility was only partial the unnatural state of the hybrids was further manifest by the union of male and female "seed materials" in differing proportions. Unions in equal proportions gave rise to F2 hybrids like their F1 parents; unions in unequal proportions gave rise to F2 hybrids closely resembling one or other of the originating species—maternal and paternal reversions. The strange motley of forms in the F2 generation was thus the direct result of tampering with nature. It was also clear evidence that no hybrids breed true and the conclusion was inescapable that fresh species intermediate between existing species cannot be formed in this way as Linnaeus believed. With great relief Koelreuter rested assured that the doctrine of the fixity of species was not refuted by his experiments.

He arrived at an alchemical explanation of the difference between F1 and F2 hybrids by analogy with salt formation. It was known at the time that when an acid and an alkali react together a salt is formed which is neither alkaline nor acid but neutral. And Koelreuter believed that when fertilisation takes place the female "seed material" unites with the male "seed material" to form a "compound material" from which the embryo is formed. He further held that as plants grow they "aim at liberating little by little, the one compound material out of which they are formed, and dividing it into the two original ground materials . . ."[15] These become concentrated in the ovules and the pollen grains and when normal fertilisation ensues the two seed materials combine again in equal proportions to form an intermediate product. In the reproduction of hybrids, however, this rule of equality is broken. Male and female "seed materials" combine in varying proportions thus producing a variety of offspring.

Several of his F2 and back-crossed populations showed a wide range of variability and so he concluded that the number of possible combinations between the two "seed materials" was very large, a state of affairs which he believed obtains also in all chemical combinations. Indeed his picture of chemical and hybrid combinations was a glorified sort of alchemical cookery according to which any variation in the quantities of the ingredients was expressed in the characteristics of the product. Thus he believed that where F1 hybrids are not intermediate between the parental species it is because a "tincture" of pollen from the mother species has acted in fertilisation alongside the foreign pollen of the paternal species. He called such products "half-hybrids" to distinguish them from true F1 hybrids.

Returning now to the F2 generation, Koelreuter recognised that in some cases—notably in hybrid pinks—the F2 hybrids are commonly of three types: those like the grandmother species, those like the F1 parents and those like the grandfather species. Broadly speaking, these are the three classes of hybrid offspring which Mendel later found to occur in the proportions 1 : 2 : 1 (for a cross involving one pair of contrasted characters). We can say, then, that Koelreuter discovered in purely qualitative terms the three segregating classes, but we should remember that by no means all his results could be fitted into these groups. Thus in the hybridisation of the marvel of Peru (*Mirabilis jalapa*) he found, as well as reversions to earlier characters, completely fresh characters, so he advised those who wished to obtain new varieties to self-pollinate hybrids generation after generation; an endless number of varieties would result.[16] The colour of his *Mirabilis* hybrids was also very varied and could not be fitted into clear-cut categories with ease. Koelreuter was dealing with a case of complex inheritance such as Mendel came up against when he studied the inheritance of flower and seed colour in the beans (*Phaseolus multiflorus* and *P. nanus*). "With regard to the colour characteristics", Mendel said, "it certainly appears difficult to perceive a substantial agreement."[17] And in a letter he wrote to Naegeli in 1870 he referred to experiments with varieties of the stock (*Matthiola annua* and *M. glabra*) which had lasted six years but which failed to give a substantial agreement with his results from *Pisum*.[18] Mendel, it is true, believed that these "enigmatical results" might be explained by postulating the action of several factors (he called them colour characters not factors), but Koelreuter, who had never seen a clear case of segregation, regarded these results with *Mirabilis* as further evidence that the natural laws of inheritance are not obeyed by the descendants of F1 hybrids. To him the important feature of the F2 generation was the presence in it of forms closely resembling the original parental species. From this fact he concluded that hybrid descendants revert sooner or later to one or other of the original species, and they do not form new species.

Mendel discussed this reversion of hybrid progeny and he explained it in terms of the ratios which he had established for *Pisum*. "The observations made by Gaertner, Koelreuter and others", he said, "that hybrids are inclined to revert to the parental forms, is also confirmed by the experiments described."[19] He showed that if monohybrids (i.e. hybrids produced from parents which differ in one character only) are self-pollinated for ten generations there would be 1,023 offspring identical with the original mother species, 1,023 identical with the original father species to every 2 hybrids. On the other hand, if several character-differences are involved the proportion of complete reversions is much smaller and continuing variability is evident. If, says Mendel,

> . . . the two original stocks differ in seven characters, and 100 or 200 plants were raised from the seeds of their hybrids to determine the grade of the relationship of the offspring, we can easily see how uncertain the decision must become, since for seven differentiating

> characters the combination series contains 16,384 individuals under 2,187 various forms; now one and then another relationship could assert its predominance, just according as chance presented this or that form to the observer in a majority of cases.[20]

In both cases the result of continued selfing is the elimination of hybrid types (heterozygotes) and the production of true-breeding types (homozygotes). The selfing of hybrids derived from multi-factorial crosses also yields new combinations of characters. Novel, true-breeding forms can thus arise in the process of hybridisation, and Mendel realised that these forms are significant for evolution. At the same time it must be pointed out that, in general, new species rarely arise directly and solely from hybrids. Genetic and geographic isolation usually play a part in the process of species formation which is not a sudden event excepting where polyploidy is involved.

Koelreuter realised that new varieties arise as a result of the selfing of hybrids, but he denied the relevance of this fact to the question of the origin of species. He was, of course, unaware of the effects of isolation and polyploidy, so the evidence for the origin of species from hybrids seemed far less convincing in his day than it does in ours. This state of affairs, together with the conclusions of Koelreuter and Gaertner, were later to draw Darwin's attention away from hybrids as species formers, which was unfortunate. But in the early part of the nineteenth century it served to make biologists aware of the exaggerated nature of Linnaeus' claims for the evolutionary role of hybrids.

Preformation refuted

The part of Koelreuter's work which had the greatest impact was his refutation of another naïve but widely held theory—the theory of preformation. According to this theory the development of an organism is no more

JOSEPH GOTTLIEB KOELREUTER
Med. Doct. Mitglied der
Kaiserl. Academie
der Wissenschaften
zu St. Petersburg
Geb. zu Sulz am Nekar den 27. Apr. 1733
Gest. zu Carlsruh den 12 Novemb. 1806.
Dr. Kölreuter

1 *Joseph Koelreuter*

2 *The intermediacy of hybrids:*

a *A reconstruction of the flowers from measurements given by Koelreuter.*

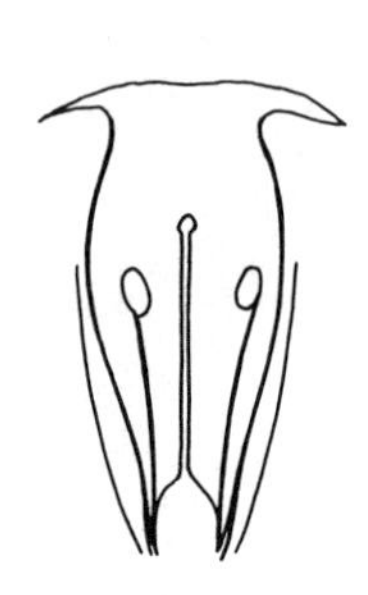

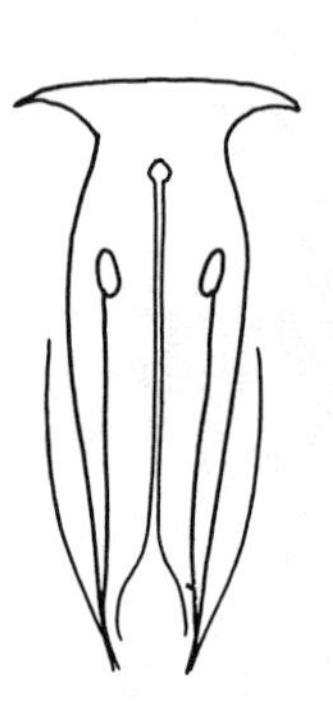

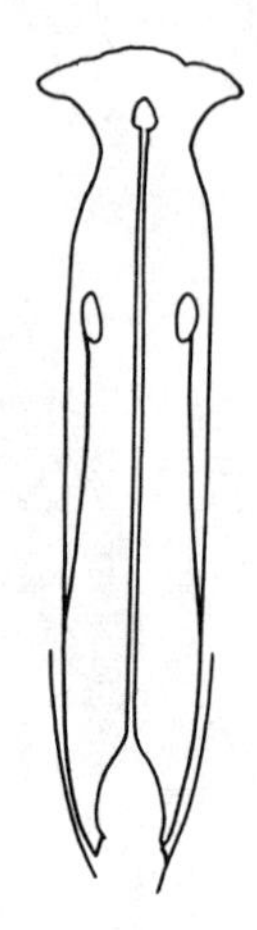

Nicotiana rustica   Nicotiana rustica × paniculata   Nicotiana paniculata

b *Intermediacy at the cellular level, based on I. J. Macfarlane's illustrations of* 1895

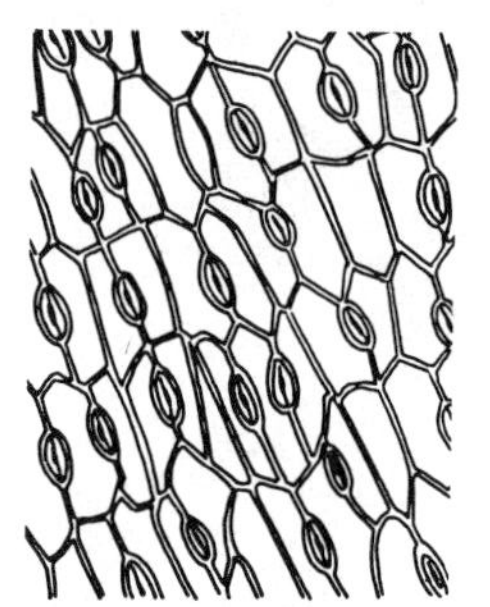

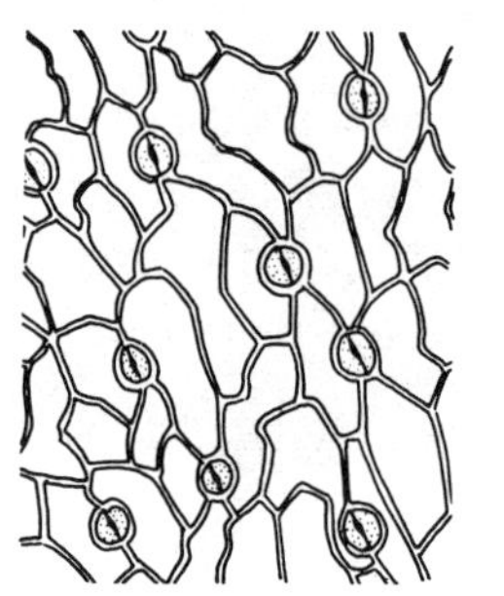

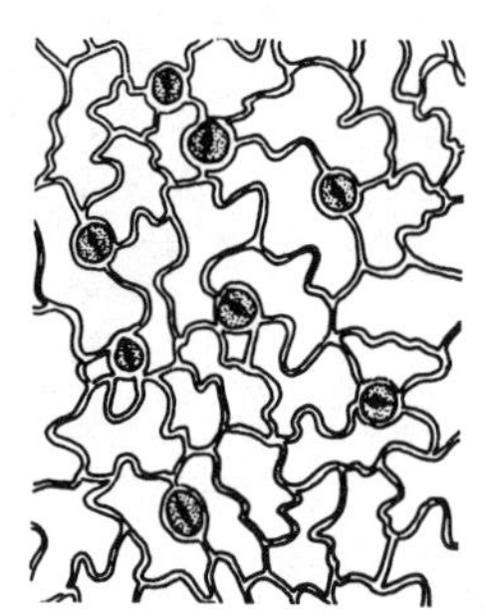

Philesia buxifolia   P. buxifolia × L. rosea   Lapageria rosea

3a *The house of the Gaertner family as it was in the eighteenth century*

than the unfolding of that which is already present in miniature. Every organism must therefore contain in its reproductive organs an infinite series representing all its future descendants. This simple and picturesque theory explained everything except which sex is the proud possessor of the species' future. The ovists said it was the female, the spermists said it was the male. The battle between these rival factions was raging in the time of Buffon and Linnaeus who, as we have seen, resolved the matter by denying preformation and proposing the two-layer theory of heredity. Koelreuter decided to test both preformation and the two-layer theory by observation and experiment. He left on one side the facts concerning the mule and hinny and dealt exclusively with plants. What he obtained was a threefold refutation of both preformation and the two-layer theory, based on the intermediacy of hybrids, the identity of reciprocal crosses and the transmutation of one species into another by successive cross-pollinations.

His first example of intermediacy came in 1761 when his hybrid tobaccos (*Nicotiana rustica* × *N. paniculata*) came into flower. The spread of the branches, the position and colour of the flowers, and the individual parts of the flower excepting only the anthers were all intermediate. In 1763 he gave the measurements of thirteen quantitative floral characteristics for this hybrid and for its parents which seemed to prove his point adequately. A large proportion of the hybrid values for these thirteen character differences approached very closely to the parental means. Naturally he expected the same of all other hybrids. When he crossed double-flowered pinks with single pinks the offspring were all double. This seems to have been the only striking exception which he encountered to the rule of intermediacy. The inference, he said, was clear—both sexes contribute equally to the progeny, neither the male nor the female seed acts exclusively, but both act together.[21]

The intermediacy of hybrids is not, of course, by itself sufficient proof that heredity is controlled by both parents, and to an equal degree. Such intermediacy could result from the weakening of the genetic force of the one species, assuming for the moment that it alone carries the heritage of the species, by the unnatural union with the other. But Koelreuter devised another test; he carried out as many as possible of his crosses in both directions. Thus the cross *Nicotiana paniculata* × *N. rustica* which failed in 1761 succeeded in 1762. He raised eight plants which were all so like their reciprocals, *N. rustica* × *N. paniculata*, "that often", he said, "I was myself unable to distinguish between the two kinds, if I did not look at the numbers.[22] This circumstance," he concluded, "confirms afresh the theory of reproduction by means of both kinds of seed."[23]

His final and most convincing proof of biparental heredity was his famous species transmutations. We have seen that Koelreuter noted the tendency of hybrid offspring to revert to the original species. He thought that if he repeatedly pollinated the hybrid offspring, generation after generation with the same parental species he would eventually obtain offspring identical with that species. His first success was the conversion of *Nicotiana rustica* to *N. paniculata* in 1763. He was so amazed that he declared that "I did not know whether it would be a very much more remarkable thing if a cat were seen to emerge in the form of a lion".[24]

Later he succeeded in transmuting three species of *Dianthus* and one species of *Mirabilis*. If he had had the time and money, he would have attempted the transmutation of the canary into the goldfinch, but even without such experiments Koelreuter felt justified in denying preformation.

Mendel thought these species transmutations merited special mention so he devoted three-and-a-half pages of his famous paper on *Pisum* to explaining why some of them took longer than others; in particular why the transmutation often took longer in one direction than in the other. He showed that it depends on the number of experimental plants and the number of differentiating characters involved. Gaertner thought that the differences in the two directions was evidence that the two natures in hybrids are not in perfect equilibrium as Koelreuter had asserted. But Mendel, having demonstrated the true reason was able to say that "Koelreuter does not merit this criticism".[25] And we can add that Koelreuter merits nothing but praise, for by his pioneer work in experimental hybridisation he overthrew all the long-established beliefs in heredity as merely an expression of the relative sexual forces and the more recent suggestions of arbitrary divisions between male and female determination of characters such as Linnaeus and Buffon were content to accept. At last the way to the solution of the vexed question of the laws governing heredity had been shown. It lay in the experimental study of plant hybrids.

NOTES

(1) Le Comte de Buffon, G.L.L. 1749–1804. *Histoire Naturelle* . . . 44 Tom. Paris. Suppl. 3. 1776. *Servant de suite à l'histoire des animaux quadrupèdes. p.* 12.

(2) Thomas Fairchild is said to have produced a hybrid between the carnation (*Dianthus caryophyllus*) and the sweet william (*D. barbatus*) before 1719.

(3) Siegesbeck, J. G. 1737. *Botanosophiae verioris brevis Sciagraphia . . . Systema Plantarum Sexuale* . . . Petropoli. p. 49, cited in: Stoever, D. H. 1794. *The Life of Sir Charles Linnaeus . . . translated by Joseph Trapp*. London. p. 121.

(4) Afzelius, A. 1923. *Egenhändiga anteckningar af C. Linnaeus* . . . Upsaliae, p. 90. Cited in: Hagberg, K. 1952. *Carl Linnaeus, translated by Alan Blair*. London. pp. 208–209.

(5) Haartman, J. 1764. "Plantae Hybridae quas sub praesidio D: n. Doct. Caroli Linnaei, . . . Upsaliae 1751 . . ." Amoenitates Academicae, *3*, 28–62.

(6) Stolberg, F. L. Count von. 1796–1797. *Travels through Germany, Switzerland Italy and Sicily translated by Thomas Holcroft*. 2 vols. London. vol. i, p. 33.

(7) Rose, H. 1775. *The elements of botany . . . being a translation of the Philosophia Botanica, and other treatises of the celebrated Linnaeus* . . . London. p. 16.

(8) Linnaeus, C. 1786. *A Dissertation on the sexes of plants translated by James Edward Smith*. London. p. 55.

(9) Koelreuter, J. G. 1761–1766. *Vorläufige Nachricht von einigen das Geschlecht der Pflanzen betreffenden Versuchen und Beobachtungen, nebst Fortsetzungen 1, 2*

und 3. Lipsiae. Reprinted in: *Ostwald's Klassiker der exakten Wissenschaften*, No. 41. Leipzig. 1893. p. 43. (Referred to in future as "OK".)
(10) OK, p. 44.
(11) OK, p. 45.
(12) OK, p. 31.
(13) OK, p. 30. (The second time he crossed *Nicotiana rustica* × *N. paniculata* the hybrids were closer to *N. paniculata*; see: OK, pp. 52–53.
(14) OK, p. 44.
(15) OK, p. 43.
(16) Koelreuter, J. G. 1794. "Mirabilium Jalapparum hybridarum continuata descriptio". Nova Acta Academia Scientiarum Imperialis Petropolitania, *12*, p. 398.
(17) Mendel, G. 1956. *Experiments in plant-hybridisation*. Cambridge, Mass. (Reprint of Professor Bateson's translation of 1901), p. 30. (Referred to in future as "MR1".)
18) Mendel, G. 1950. "Gregor Mendel's letters to Carl Naegeli. 1866–1873. Translated by L. K. Piternick and G. Piternick." (Supplement to Genetics: "The Birth of Genetics") Genetics, *35*, No. 5, pt. 2, p. 25. (Referred to in future as "MR2".)
(19) MR1, p. 14.
(20) MR1, p. 34.
(21) OK, p. 31.
(22) OK, p. 45.
(23) OK, p. 45.
(24) OK, p. 66.
(25) MR1, p. 39.

SUGGESTIONS FOR FURTHER READING

Behrens, J. 1895. "Joseph Gottlieb Koelreuter. Ein Karlsruhe Botaniker des 18 Jahrhunderts." Verhandlungen des naturwissenschaftlichen Vereins in Karlsruhe, *11*, 268–320.

Glass, B. 1959. "Heredity and variation in the 18th century concept of species." In: *Forerunners of Darwin 1745–1859*. Edited by Glass, Temkin and Straus. Johns Hopkins Press, Baltimore. Chapter vi.

Roberts, H. F. 1929. *Plant Hybridization before Mendel*. Princeton University Press, Princeton.

Sachs, J. von. 1890. *History of Botany (1530–1860)* trans. H. and E. F. Garnsey, revised by Bayley Balfour. Oxford University Press, Oxford.

Zirkle, C. 1935. *The beginnings of plant hybridisation*. (Morris Arboretum Monograph No. 1.) University of Philadelphia Press, Philadelphia.

Chapter Two

Hybridisation before Mendel (1787-1849)

In 1761 Koelreuter expressed the view that as a result of his experiments "even the most stubborn of all doubters of the sexuality of plants would be completely convinced", and if he were not, "it would astonish me as greatly if I heard someone on a clear midday maintain that it was night".[1] Little did he know that both his work and the fact of plant sexuality would be attacked long after his death. The principal protagonists in this story were Friedrich Schelver (1778–1832), a professor in Heidelberg, and August Henschel (1790–1856), a medical practitioner and university tutor in Breslau. Their chief reasons for opposing Koelreuter seem to have been that they respected tradition and distrusted experiments. Both men preferred the doctrine of asexuality in the plant kingdom as taught by the ancients to the opposite doctrine of plant sexuality as held by the moderns. And Henschel gave a very different interpretation to Koelreuter's experiments, from that of their author.

His interpretation can be described as follows. If flowers are castrated and dusted with pollen from a plant of another species, you cannot expect them to behave normally. These disturbances are in themselves sufficient to reduce the fertility of the plant and cause deviations from the parental type in the offspring. Had Koelreuter not himself admitted that amongst the progeny of hybrids there were monstrosities (*Missgeburten*) and that all such progeny showed a high degree of variation? Henschel's conclusion was that what Koelreuter regarded as hybrid characteristics were no other than the characteristics of varieties, degenerations and monstrosities produced by growing plants in artificial conditions and tampering with their flowers.

Henschel's own view of the process of seed formation was based on the principles of the "Naturephilosophers". This school of philosophy was headed by such men as Goethe, Hegel, Oken and Nees von Esenbeck. They viewed nature in anthropomorphic and spiritual terms; they considered all nature as a unity in which the whole is expressed in the parts, and they treated the individual organism and its constituent parts in like manner. Hence we find Henschel reasoning that since the pollen is only a very small part of the plant it cannot transmit the whole nature of the plant to another plant. If another plant is needed it must be the whole plant, and surely the mere proximity of one plant to another should suffice? He preferred to consider the formation of pollen as the final act in the progressive liberation of the spiritual nature of the plant from its material nature. Having completed this act, and only then, is the plant in a fit state to form seeds.

There were, naturally, many factual obstacles in the way of Henschel's arguments. These he either clambered over in a rather clumsy fashion or skirted around them, but such was the popularity in Germany of this "Nature-philosophy" that Henschel's views won a wide acceptance amongst those who were not addicted to experimental science, and from those who were came a request for further experiments on the lines of Koelreuter's. Two prizes for an account of such experiments were offered, one in 1822 and another in 1830. The winner of the first prize was Dr. A. F. Wiegmann (1771–1853) who was known to many as the editor of *Wiegmann's Archiv für*

Naturgeschichte. He grew his plants in the open soil so that his results could not be attributed to pot culture—Henschel had criticised Koelreuter for growing his plants in pots. The winner of the second prize was Carl Friedrich von Gaertner (1772–1850). His work was on a far larger scale than Wiegmann's and is more important in the history of genetics. He was the direct successor to Koelreuter and the direct precursor of Mendel.

Gaertner's Life

Carl spent the first fifteen years of his life in Calw, the home of his father, Joseph Gaertner (1732–1791), the world-famous botanist and friend of Koelreuter. He was sent to the monastery school in the little town of Bebenhausen near Tübingen, where he received excellent instruction in science. His father took a lively interest in his son's progress and by his informative letters fostered Carl's interest in the sciences. From Bebenhausen Carl went to the Carlsakademie in Stuttgart to study medicine. His studies which were chiefly biochemical led to the degree of Doctor of Medicine in 1796.

After Joseph's death in 1791 Carl worked on what remained to be published of his father's famous three-volume book *De Fructibus et Seminibus Plantarum*. The botanical studies which he was thus led to undertake found a natural meeting-point with his earlier biochemical work in plant physiology and on this subject he decided to write his *magnum opus*, using Haller's *Physiology* as his model. He had already filled twenty-six octavo volumes with closely-written notes and quotations before he decided to limit his task to the subjects of fertilisation and hybridisation in plants. Two events which, it would seem, helped to bring him to this decision were the controversies which occurred over the sexuality of plants after Henschel's book appeared in 1820, and the supposed direct colour action of foreign pollen on the coats of the resulting seeds. John Goss had described experiments which demonstrated such action and Alexander Seton had described similar experiments which did not demonstrate it. Thus we find Gaertner in 1824 crossing varieties of maize with different seed colours, and in 1825 repeating some of Koelreuter's experiments with tobaccos. He continued hybridising plants for another twenty years. In the 1830s and 40s he contributed many papers on his experiments to the *Regensburg Flora* and lectured at botanical conferences in Heidelberg, Erlangen and Stuttgart.

In 1830 the Dutch Academy of Sciences offered a prize for an answer to the question:

> What does experience teach regarding the production of new species and varieties, through the artificial fertilisation of flowers of the one with the pollen of the other and what economic and ornamental plants can be produced and multiplied in this way?

No entries were received, so the closing date was extended to 1836. Gaertner first heard of this prize in 1835. Instead of an essay he sent a brief résumé of his work to the secretary of the Academy. The commission for the prize was delighted with this and granted him further time in which to complete a full report. Gaertner sent this in 1837 and was awarded the prize. After subsequent revision and translation into Dutch his report was published in 1838. The essay received its final revision and enlargement in 1849 when it was published in German for the first time, under the title *Versuche und Beobachtungen über die Bastarderzeugung im Pflanzenreich* (Experiments and Observations upon hybridisation in the plant kingdom). This book contains reports of nearly 10,000 separate experiments among 700 species which yielded 250 different hybrids. Darwin said of this book that "it contains more valuable matter than all other writers put together, and would do great service if better known".[2] Mendel remarked that it contained records of "very valuable observations",[3] and from the frequent underlining in his copy, now preserved at the Mendel Museum, we can be sure that he read it from cover to cover.

The book marked the culmination of Gaertner's work, but he did not live to hear the praise, for he died soon after it appeared. Though he failed to achieve international fame he was renowned in Calw and the surrounding district of Swabia. In 1846 the people of Calw celebrated the fiftieth anniversary of his award of a doctorate. The Margrave of Württemberg made him a Knight of the order of the Crown and the citizens of Calw accorded him the freedom of their city. At a meeting of naturalists in Heilbronn a year later a poem was recited in his honour which translates roughly as follows:

> To Kerr's toast of Gaertner I add this:
> Hail Gaertner! Hail the 1st of May,
> Then listen! On this day he was born.
> This (fact) and his name Gaertner tells us freely
> That God declared him for the world of flowers.

Gaertner's personality and the circumstances of his life were very different from those of Koelreuter. Unlike the latter, Gaertner was not arrogant, irascible and difficult to get on with, but was unassuming and good-natured. He began with two advantages over his predecessor, Koelreuter; he was the son of a world-famous botanist, and had sufficient private means to enable him to devote time and money to his hybridisation experiments over a period of twenty years. Also, unlike Koelreuter, he had at Calw a fair-sized garden of his own in which he could pursue his experiments unmolested by officious gardeners. This same garden in Calw had been used by Koelreuter when he was the guest of the Gaertner family in 1762.

It would be of little value to review all the work which Gaertner discussed in the *Bastarderzeugung im Pflanzenreich*, for he was an encyclopaedic writer and his book is lumbered with a vast number of references to every conceivable writer on the subject. There are thirty-eight chapters, some of which contain between one and two hundred references each. His writing is repetitive and lacks the brilliance of Koelreuter, both in style and content. But he was a most patient and careful worker and his book was the only exhaustive treatise on the subject at the time. We will only concern ourselves here with his accounts of the crossing of peas, maize and tobacco.

Peas

For several centuries before the time of Gaertner there had been reports of different coloured seeds in the same pea pod, of different coloured maize grains in the same ear, and of different coloured fruits on the same apple tree, peach tree and so on. Dr. Conway Zirkle has traced records of this sort back to the sixteenth century. But the first person to investigate the matter by careful experiments was the British horticulturist and experimentalist Thomas Andrew Knight (1759–1838). He chose the edible pea as his material because of "the numerous varieties of strictly permanent habits of the pea, its annual life, and the distinct character in form, size and colour of many of its varieties".[4] This statement made in 1823 reminds one of Mendel's remark about *Pisum* forty-two years later. He said, "Some thoroughly distinct forms of this genus possess characters which are constant, and easily and certainly recognisable . . ."[5]

Knight's example was followed by Alexander Seton, a regular contributor to the *Transactions of the Horticultural Society of London*, and by John Goss, a Devonshire man whose interest was stimulated by reading an account in the Society's *Transactions* of nectarines and peaches produced on the same branch. The results of the experiments of these three men was the discovery of dominance of yellow cotyledon colour over green, of segregation into yellow seeds and green seeds in the F2 generation and of the true breeding character of the recessive green-seeded segregates. But on the question of the direct action of foreign pollen Knight and Seton's experiments were negative whilst Goss' were positive. In fact, it seems that Goss misinterpreted his results. His seeds must have had transparent coats whereas Seton's and Knight's had opaque coats. Now it is well known that the seed coat is produced by the mother plant before fertilisation takes place; Knight knew this in 1787 when he was hybridising peas. So the only part of the seed of a pea which has a hybrid constitution is the embryo (we are speaking of the seeds produced from the cross-fertilisation and not of subsequent generations). In peas the most conspicuous characters of the embryo are the round or wrinkled surface and the green or yellow colour of the seed leaves or cotyledons. What Goss took for the colour of the seed coats was in fact that of the cotyledons underneath them. Hence there is no direct action of foreign pollen on the seed coats, or on the fruits for that matter, only on the embryo. Thomas Knight put forward this explanation in 1823 but he won little support for it. Gaertner therefore crossed pea varieties with differently coloured seeds in 1829. Some of his results were like those of Goss, and others similar to those of Seton and Knight. He did not grow the second generation as far as we know, for he was interested only in the question of the direct action of pollen on the F1 seeds.

Maize

Maize seeds do not have large fleshy cotyledons as do pea seeds. They have instead a nutritive tissue called endosperm. This, we now know, originates from a nucleus of the pollen grain and two nuclei of the mother plant. The effects of hybridisation can therefore be seen immediately in this tissue. Gaertner had maintained constant for several years a short variety of maize which bore small yellow seeds, and three tall varieties which bore large, brown, red and red-striped seeds respectively. He crossed the short plants with the three tall varieties in

1824. Only pollen from the red-striped variety was effective, and he could discern no immediate change of colour in the resulting seeds. But the F2 seeds produced from this cross were of several colours. In two ears he counted the greyish-red and reddish-grey seeds and found 224 to 64 in one and 104 to 39 in the other.[6] This gives the ratio 3·18 to 1, and since yellow endosperm colour is dominant to red, Gaertner may well have observed clear Mendelian segregation. He also recorded the fact that the plants grown from the darker of the grey seeds yielded a mixture of seeds in the following proportions: pure yellow—about $\frac{1}{4}$, yellow and grey streaked—nearly $\frac{1}{8}$, reddish-grey—$\frac{1}{12}$, dark reddish-grey and brownish-red—$\frac{1}{2}$.[7] These results are difficult to interpret, but are what one would expect with many varieties of maize, especially striped varieties. The difficulty with maize is that the seeds are all invested in a membranous fruit wall or pericarp which is sometimes transparent and at others coloured. The striped appearance of the grains is due to the striping of the pericarp and this character of the fruit is subject to considerable variation over the grains in one and the same ear. Nevertheless, there are varieties of maize for which it is easy to obtain clear Mendelian segregation, and Hugo de Vries, who was the first to discover Mendelian segregation after Mendel, succeeded with maize in 1898. (See Chapter 6.)

Goss, Seton, Knight and Gaertner all observed dominance and segregation. What we want to know is, what did they make of these discoveries? First, they did not arrive at a clear expression of the latency of characters when masked by others. Second, not one of them suggested that some segregating mechanism takes place in the reproductive process. The general idea was that family variability was due to the action of two different males. Knight thought this might explain the variability in the F2 generation. Gaertner rightly condemned this answer, but his own was too vague to be of any value. In 1828 he spoke of the "impulse to colour seeds" being "distributed at random"[6] and in 1849, of the capacity of the plants to bring forth either of two "concurring factors".[9] Thus the basic idea was that there is no segregation. The F2 seeds, like the F1 seeds, retain a double constitution and reversion is due to selective development of one factor at the expense of the other. The two quite distinct processes of dominance and segregation were thus confused and combined in these explanations.

Tobacco

Gaertner, like Koelreuter, carried out a great number of crosses in the tobacco genus. He classified the resulting hybrids into three groups: intermediate, commingled and biased. Koelreuter had regarded all true hybrids as intermediate but Gaertner rightly held that by far the greater number belong to his second class—commingled hybrids —in which some parts of the hybrid approach closer to the pollen parent and other parts to the maternal parent.

Hybrids of the biased type are far less common, but Gaertner described some very striking examples. He called them "biased" (*decidirten*) because they were so like one of the parental species that at first sight they appeared not to be hybrids at all. Thus when he crossed *Nicotiana suaveolens* and *N. suaveolens* var. *vincaeflora* with *N. langsdorffii* the latter species could be detected in the resulting hybrids only in the following characters: separation of the filaments from the corolla tube, bluish colour of anthers, greenish colour and curving shape of the corolla tube.[10] When crossed with *N. tabacum* var. *macrophylla*, on the other hand, *N. suaveolens* was not detected by Gaertner in the resulting hybrids. These facts must have strengthened his conviction that the species acts as a whole. Indeed, when in 1916 Thomas Goodspeed and R. E. Clausen produced biased hybrids between *N. tabacum* and *N. sylvestris* they felt it necessary to postulate an "hereditary reaction system" for each species which acted as a whole.[11]

Gaertner did not know what conclusion to draw from these very diverse results. He went as far as to ask whether it is the "total nature of the species" or that of the separate organs which determines the direction and form of the resulting hybrids,[12] but in the end he clung to the generally held view that it is the former alone. With this conclusion it was, of course, impossible for him to arrive at the Mendelian explanation of hybridisation.*

This brings us to the question of the suitability of tobacco species for genetical studies. In the eyes of Koelreuter and Gaertner they were eminently suitable since they are easy to grow, easy to castrate and cross, and they set numerous seeds. Moreover, the many species of tobacco cross with surprising facility, despite the fact that there is a wide range of chromosome numbers within the genus. What they did not know, however, is that these differences in chromosome number introduced a complicating factor into their experiments. Unwittingly they were observing the effects of polyploidy as well as of simple hybridisation. Thus the biased tobacco hybrids which Gaertner produced were due not to the dominating action of the species as a whole but to the double dose of chromosomes supplied by one parent. The germ cells of *N. suaveolens* contain 16 chromosomes each, those of *N. tabacum* 24, but those of *N. langsdorffii* have only 9.

Koelreuter's famous first hybrid *N. rustica* × *N. paniculata* also involves a difference in chromosome number. *N. paniculata* is a diploid whose germ cells have 12 chromosomes each, whilst *N. rustica* is a tetraploid whose germ cells have 24 chromosomes each. The resulting hybrid is therefore a triploid. Triploids, unlike diploids, do not show normal Mendelian segregation in the F2 generation because the production of pollen and ovules

* But it would not have prevented him from arriving at Naudin's theory of specific segregation.

is subject to irregularities. Mendel's theory only applies to those cases in which the germ cells contribute an equal amount of genetic material to the hybrid. In the cases under discussion the two germ cells contain different numbers of chromosomes, hence they do not contribute equally. Mendel also assumed "that the various kinds of egg and pollen cells were formed in the hybrid on the average in equal numbers". This does not apply in the case of triploids where some associations of chromosomes are non-viable and fail to form germ cells, or only form ineffectual cells, whilst other associations form cells effectual in fertilisation. In the case of the hybrid *N. rustica* × *paniculata* only about one in a thousand pollen grains are viable according to Edward East,[12] and the majority of these have nearly as many chromosomes as do the germ cells of *N. rustica* (i.e. 24). The effect of this non-uniformity in the viability of germ cells is thus to intensify the reversion in the F2 generation to the parent with the higher chromosome number.

Koelreuter and Gaertner found the marked infertility and tendency to reversion of triploid hybrids agreed well with their view that species remain distinct and unchanging. The results which they obtained from the crossing of varieties of garden pink, marvel of Peru and thorn apple, however, were very different. In these cases fertility suffered little if any diminution and hybrid variability was in general very persistent. Hence to them it seemed abundantly clear that between species and varieties there is a gulf which cannot be bridged. This conclusion would not of course be accepted today, but there was an element of truth in it since the behaviour of triploid and diploid hybrids is different.

We can thus appreciate that some of the results of these early hybridisations merited separate classification, but it was unfortunate that a corollary of their distinction was the assumption that results obtained from crossing of varieties has little if any relevance to the question of how new species originate. They saw the economic importance of their results for agriculture and horticulture and that was the limit of their vision.

The Separation of Species from Varieties

The view that the products of man's art are unnatural can be traced back many centuries. It is found in the early commentaries on the Bible where the "fall" of man is believed to cover not only man's moral nature but also that of all he produces and anything in nature with which he interferes. Hybridisation, because it is unnatural, leads to the production of "degenerations", and this term was used of domesticated animals and cultivated plants irrespective of whether they were more or less valuable to man than their wild relatives. They were described as degenerate because it was widely held that they do not breed true. As early as the thirteenth century St. Augustine excluded hybrids from the Ark[13] and in the sixteenth century, when rationalists were questioning the story of the Ark because all the species of animals then known could not possibly be fitted into it, St. Augustine's exclusions were made use of by the pious in their defence of the Mosaic tradition. Thus Joshuah Sylvester (1563–1618) said:

O profane mockers! if I but exclude
Out of this Vessell a vast multitude
Of since-born mongrels, that derive their birth
From monstrous medly of *Venerian* mirth:
Fantastick Mules, and spotted Leoperds,
Of incest-heat ingendred afterwards:
So many sorts of Dogs, of Cocks, and Doves,
Since, dayly sprung from strange and mingled loves,
Wherein from time to time in various sort,
Dedalian Nature seems her to disport: . . .[14]

This defence of Biblical tradition was continued in the seventeenth and eighteenth centuries especially by the Protestants. Linnaeus, who was intended for the Lutheran

3b *Gaertner's garden. Here Koelreuter continued his hybridisation experiments during* 1762–3, *when he was the guest of Achatius Gaertner. (The latter studied medicine with Koelreuter at Tübingen, and was a cousin of Carl's father, Joseph Gaertner. Carl later used this garden for his experiments)*

4 *The Market Place, Calw, as it is today. Gaertner's house bears the name 'Paul Räuchle'. There is a tablet to Koelreuter and the Gaertners beside the door on the right*

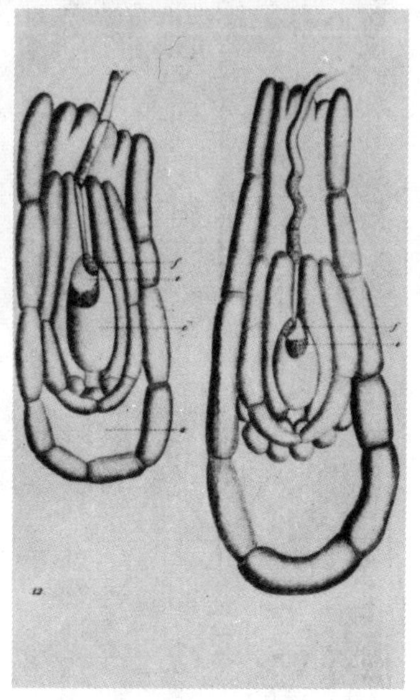

5a *Amici's illustration of the penetration of the ovary of* Orchis morio *by the pollen tube. c = nucellus, e = embryo sac, f = tip of pollen tube. From* Ann. Sci. Nat. Botanique, *sér:* 3, 7 (1847), *Plate* 10

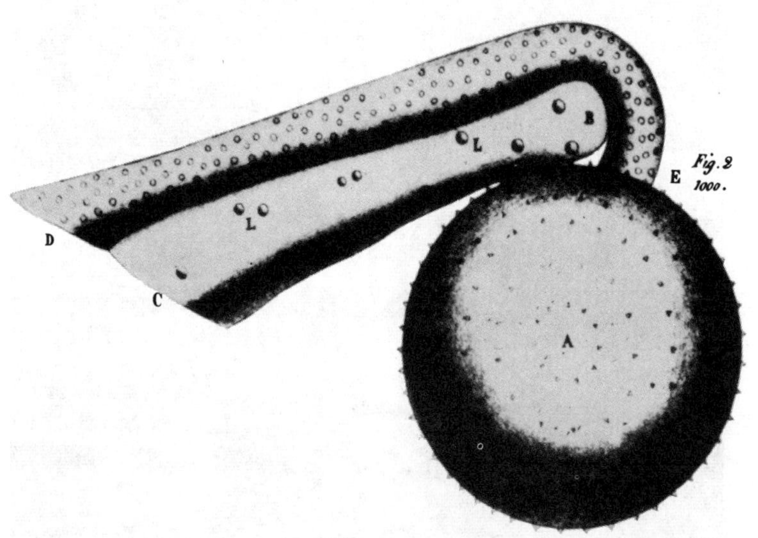

5b *Amici's illustration of a pollen grain of the oleander showing a portion of the pollen tube* (D-E) *which has grown along the stigmatic hair* (L-B). *From* Ann. Sci. Nat. *sér.* 1, 2 (1824) *Atlas, Plate* 4

Church, was steeped in Protestant theology, so it is not surprising that when he defined species and varieties in 1737 he used the theological distinction between God's perfect and unchanging world and man's imperfect and changing world. He said of natural species and garden varieties:

> . . . I distinguish the species of the Almighty Creator which are true from the abnormal varieties of the Gardener: the former I reckon of the highest importance because of their author, the latter I reject because of their authors. The former persist and have persisted from the beginning of the world, the latter, being monstrosities, can boast of but a brief life.[15]

This is one of the 324 rules of nomenclature which Linnaeus put forward in his *Critica Botanica* with the chief purpose of producing order in a confused science. This he achieved with remarkable success, but his contemporaries and successors, in following his instructions to the letter became used to a very static idea of nature. Koelreuter and Gaertner, excellent experimentalists though they were, clung to this static view and as a result failed to see the evolutionary significance of their work.

The natural species was thus the favourite experimental material. In Gaertner's work it was also the favourite unit. He thought in terms of the essence and nature of the species, its building force and "elective affinity". The latter term was introduced by Gaertner to express the power of one species to fertilise another. He expressed it quantitatively as:

$$\frac{\text{no. seeds resulting from the pollination of sp. A by sp. B}}{\text{no. ,, ,, ,, ,, ,, ,, A ,, A}}$$

To determine the elective affinity of several species he carried out a number of careful experiments, but his results are of no great importance save that they illustrate in quantitative terms a graduation in fertility between the various interspecific crosses of a genus.

The Whole versus the Parts

Gaertner's preoccupation with whole entities and natural species contrasts markedly with Mendel's analysis of unit-character inheritance in garden varieties; but their experiments are not separated by more than two decades. Gaertner's work belongs to the 1830s and 1840s, Mendel's belongs to the 1850s and 1860s.

Between 1830 and 1860, however, a radical alteration of biological thought took place which has much to do with the very different approaches of these two hybridists. When Gaertner began his experiments fertilisation was still widely believed to be a mixing of two germinal fluids and the formation of the organism out of these fluids was only conceived in the vaguest of terms. Botanical instruction in the universities was still strongly orientated towards Linnean botany and in Germany the school of Naturephilosophy was foremost in the faculty of philosophy and its teachings permeated biology. At the beginning of this chapter we saw that Goethe was a prominent member of this school. He encouraged Henschel in his attack on the experimental work of the hybridists. Goethe was also opposed to analytic treatments of organisms. He had a special dislike of those anatomists who spared no thought for the organism as a whole entity. Thus he makes Faust scoff at such men and say "he has the part in his hand, only unfortunately he lacks the living bond".[16] In 1817 in the periodical *Zur Morphologie*, which he started, he reverted to this subject again saying: "The living organism is indeed divisible into elements, but one cannot reconstitute it from these and make it live."[17]

Gaertner was educated at Tübingen and Göttingen where he imbibed the essentials of the Naturephilosopher's view of organisms. Consequently he had little sympathy with the aims of the early cytologists. These anatomists with their microscopes, he declared, "cannot possibly yield sufficient information about a phenomenon which is itself life and which is annihilated with the destruction of the integration of the parts",[18] and the suggestion that the embryo has its basis in a pre-existing foundation cell or nucleus was in his opinion "purely metaphysical", for in reality it "must proceed from a fluid".[19] Nor did he accept the truth of Amici's discovery of pollen tubes growing from the pollen grains through the stigma and down to the embryo sacs there effecting fertilisation.

Gaertner's attitude to the new cytology was the natural orthodoxy of an old man who had been brought up on the teaching of the Naturephilosophers. This school of thought suffered a mortal blow when the fiery professor from Jena, Matthias Schleiden, led a spirited and vehement campaign against it. In his famous book *Die Botanik als inductive Wissenschaft oder Grundzüge der wissenschaftlichen Botanik* . . . 1842–43 (Botany as an inductive science or Principles of scientific botany . . .) he devoted 131 pages to a discussion of scientific method. Here, in the words of Ferdinand Cohn he rejected

> . . . with fiery eloquence the sterile speculations of nature-philosophy, raised instead a banner devised on Bacon's inductive and Kant's critical method. The text handled with keen and trenchant ridicule the scholastic treatment of botany then usual in Germany. It scoffed at florists and systematists as gleaners of so much hay. It established two maxims as supreme watchwords of science: the study of developmental history as the key to all morphology, and the study of the cell's structure and life as the key to plant physiology.[20]

Many of Schleiden's statements were outrageous and his personal attacks unjust. There were also glaring errors in his interpretation of what he saw under the microscope.

His most important error concerned the function of the pollen tube. Like Amici he saw it grow down through the tissue of the style and reach the embryo sac, but because the embryo formed at the point where the tip of the tube had indented the sac he inferred that the embryo is the transformed tip of the pollen tube. So he concluded that there is no fusion of sexual elements, that the mother plant acts only as a nurse to the embryo whose form is determined by the male alone. Thus for the third and last time plant sexuality was denied. Mono-parental heredity was reasserted. To those who advanced the facts of hybridity as evidence against his theory he replied in haughty terms. It was easy, he said, to explain how both embryo sac and pollen grain can imprint their type on the offspring, the former by way of nutriment and the latter by its contents. But he did not consider that any such explanation was necessary, for to him "as to any man of normal intelligence it appeared that hybridisation was entirely one-sided, whether *a* comes to *b* or *b* to *a*, so long as they come together".[21]

Schleiden was so tenacious and unquenchable that a whole team of cytologists had to work for years before he could be brought to his knees. Finally, in 1856, he was prevailed upon to recant his former statements when his own student Radlkofer demonstrated preparations of the embryo sac at various stages in the fertilisation process which left no doubt whatever that Schleiden was wrong. In the same year Nathaniel Pringsheim, working with the freshwater alga *Oedogonium*, observed the passage of the contents of the male agent of fertilisation, the antherozoon, into the female agent, the oogonium. He drew four most important conclusions from this observation. They were first that there is a mixing of the contents of the two germ cells, second that the foundation cell of the new organism is the direct result of fertilisation, third that the antherozoon does not form a particular part of this foundation cell, and fourth that a single antherozoon suffices for the accomplishment of the sexual act.[22] The first of these conclusions was a denunciation of Schleiden's theory, the second of theories of preformation, the third of the two-layer theory of Buffon and Linnaeus which had been resurrected by Rolando, and the fourth of the widely held belief that more than one antherozoon or spermatozoon is necessary to achieve fertilisation of one egg.

Gaertner died in 1850, six years before Pringsheim made his important discovery. But although he did not live to read about it he already knew from his experiments in hybridisation that the "double paternity" of Knight and the "tinctures" and "half-hybrids" of Koelreuter are impossible. He also knew that the products of sexual reproduction are determined by definite laws. Little did he know that a successful search for these unknown laws was soon to be made with the aid of the cells and nuclei which he despised and the peas which he looked upon as mere garden varieties to which natural laws do not apply.

NOTES

(1) OK, p. 6. Cited by H. F. Roberts, 1929, p. 160.

(2) Darwin, F. (Ed.) 1903. *More letters of Charles Darwin. A record of his work in a series of hitherto unpublished letters.* 2 vols. London. Vol. ii, pp. 339–340. (Referred to in future as "MLD".)

(3) MR1, p. 1.

(4) Knight, T. A. 1841. *A selection from the physiological and horticultural papers, published in the Transactions of the Royal and Horticultural Societies by the late Thomas Andrew Knight Esq.*, . . . London. p. 279.

(5) MR1, p. 3.

(6) Gaertner, C. F. 1827. "Correspondenz." Flora, Jena, *1*, 80.*

(7) Gaertner, 1828. "Correspondenz." Flora, Jena, *2*, 557.*

(8) Gaertner. 1828. "Correspondenz." Flora, Jena, *2*, 559.*

(9) Gaertner. 1849. *Versuche und Beobachtungen über die Bastarderzeugung im Pflanzenreich. Mit Hinweisung auf die ähnlichen Erscheinungen im Thierreiche &c.* Stuttgart. p. 327. (Referred to in future as "Bz".)

(10) Bz, p. 256.

(11) Clausen, R. E., & Goodspeed, T. H. 1916. "Hereditary reaction-system relations—an extension of Mendelian concepts." Proc. nat. Acad. Sci., Wash., *2*, 241.

(12) East, E. M. 1928. "The genetics of the genus *Nicotiana*." Bibliographia Genetica, *4*, 280–281.

(13) Augustinus, D. A. 1620. *Of the city of God, with the learned comments of I. L. Vives*. 2nd ed. corrected and amended. London. p. 537.

(14) Sylvester, J. 1880. *The complete works of Joshuah Sylvester collected and edited by the Rev. A. B. Grosart*. Edinburgh. Vol. i, p. 136, lines 416–425.

(15) Linnaeus, C. 1938. *The Critica Botanica of Linnaeus*. Trans. Sir Arthur Hort . . . (Printed for the Ray Society.) London. p. 151.

(16) Goethe, J. W. *Faust*. Cited by Russell, E. S. 1916. *Form and function. A contribution to the history of animal morphology*. London. p. 50.

(17) Goethe. 1817. "Die Absicht eingeleitet." Zur Naturwissenschaft überhaupt, besonders zur Morphologie. Stuttgart und Tübingen. Vol. i, pt. 1, p. viii.

(18) Gaertner. 1833. "Ueber Fruchtbildung und Bastardpflanzen." Flora, Jena, *1*, 210.*

(19) Gaertner. 1844. *Beiträge zur Kenntniss der Befruchtung der vollkommeneren Gewächse*. Stuttgart. pp. 434–435.

(20) Cohn, F. 1895. "Nekrologe Nathaniel Pringsheim." Ber. dtsch. bot. Ges., *13*, (11). (The bracketed page numbers refer to the biographical section at the end of the volume.) Trans. in: Goebel, K. von. 1926. *Wilhelm Hofmeister, the work and life of a nineteenth-century botanist*. London. p. 6.

(21) Schleiden, M. J. 1839. "Ueber Bastarderzeugung und Sexualität." Arch. Naturgesch., *5*, No. 1, 254.

(22) Pringsheim, N. 1856. "Observations sur la fécondation et la génération alternante des Algues." Ann. Sci. Nat. Botanique, sér. 3, *5*, 256–257.

* Flora, Jena is the periodical which used to be called Regensburg Flora. I have used the modern name. The volume numbers refer only to the first or second volume for a given year.

SUGGESTIONS FOR FURTHER READING

Arber, A. 1946. "Goethe's Botany—The Metamorphosis of Plants (1790) and Tobler's Ode to Nature (1782) with an introduction and translations by Agnes Arber." Chronica Botanica, *10*, 63–124.

Cole, F. J. 1930. *Early theories of sexual generation*. Oxford University Press, Oxford.

Hughes, A. 1959. *A history of cytology*. Abelard-Schuman, New York and London.

Jaeger, G. 1851. *Zum Andenken an den 1 Mai 1772 geborenen den 1 September 1850 zu Calw gestorbenen Botaniker C. F. von Gaertner*. Stuttgart.

Ramsbottom, J. 1938. "Presidential Address. Linnaeus and the species concept." Proc. Linn. Soc. 150th session, 192–219.

Toulmin, S. and Goodfield, J. 1962. *The architecture of matter*. Hutchinson, London.

Chapter Three

Blending and Non-Blending Heredity: Darwin, Naudin, and Galton

It is now well known that the most unfortunate of the assumptions underlying Darwin's mechanism of evolution was that of blending heredity; i.e. that parental differences are merged in the offspring of bisexual reproduction so that variation is constantly being diminished. The basis for this assumption was the so-called intermediacy of hybrids which Koelreuter had regarded as a law for all "true" hybrids. We have seen that Koelreuter's successors disputed this law, but Darwin, although aware of the many exceptions, was still inclined to accept it and the inference of blending heredity which was commonly made from it. This was not a case of blindness to the facts but simply one of having strong reasons for regarding as significant for evolution only those characters which blend.

Darwin's Genetics

In order to understand Darwin's point of view we must go back to 1837 when he returned from the *Beagle* voyage and opened the first of his four notebooks on the transmutation of species. When he wrote these notebooks his chief aim was to find a mechanistic explanation for adaptation. Having read Lyell's *Principles of Geology* during his voyage[1] he had already absorbed that author's concept of change in geology as a gradual and continuous process. Darwin assumed that new forms which had been produced by gradual changes in a species over numerous generations would be adapted to the slow and important changes of geology and climate. Sudden deviations (macromutations) would not, he thought, show the sort of adaptation characteristic of species. Accordingly he sought for a process in nature by which these unwanted deviations could be suppressed. It was here that his discovery of the two forms of flowers in *Linum* and *Primula*, which he made in 1838 and 1839,[2] proved of such importance. Dimorphic flowers, though hermaphrodite, are cross-pollinated. Evidently cross-pollination was far more common than was generally thought. Hence, he argued, there must be some advantage in cross-pollination. Thomas Knight had already stated that "nature intended that a sexual intercourse should take place between neighbouring plants of the same species",[3] and in 1841 Darwin read Christian Konrad Sprengel's book, *Das entdeckte Geheimniss der Natur im Bau und in der Befruchtung der Blumen* (The discovered secret of nature in the structure and fertilisation of flowers), which contains his discovery of dichogamy (the stigmas and anthers in the same flower ripen at different times), of the role of insects in pollination and his statement that:

> . . . since many flowers are dioecious, and probably at least as many hermaphrodite flowers are dichogamous, nature appears not to have intended that any flower should be fertilised by its own pollen.[4]

Darwin could hardly have obtained these facts at a more opportune time. If crossing blended individual differences and quashed macromutations he saw that it might well be the process for which he had been seeking. Then, providing organisms were cross-breeding, the only deviations to persist would be those which tended always in the same direction and which were accumulated generation by generation so to constitute an adapted form. Thus, in his fourth notebook he said:

> Without sexual crossing, there would be endless changes, & hence no feature would be deeply impressed on it, & hence there could not be *improvement*, . . . it was absolutely necessary that Physical changes should act not on individuals, but on masses of individuals—so that the changes should be slow & bear relation to the whole changes of country, & not to the local changes—this could only be effected by sexes.[5]

The result of this line of reasoning was that he looked upon crossing not as an instrument for diversity but for uniformity. Consequently he had to look elsewhere for the causes of variation, and he turned to the conditions of life.

Thus Darwin fell a prey to the tempting simplicity of "soft" heredity directed to some extent by the conditions of the environment, and he did this because it provided a mechanism by which *gradual adaptation* could be effected. At the same time he dismissed all mutants as monstrosities because he failed to find, after a diligent search, "cases of monstrosities resembling normal structures in nearly allied forms, and these", he declared, "alone bear on the question".[6] Also he thought it "as improbable that any part should have been suddenly produced perfect, as that a complex machine should have been invented by man in a perfect state".[7] When he discovered the positive effect of natural selection in preserving adaptive variations in 1838 his conviction that macromutations are not significant for evolution became even stronger, for he saw that selection could preserve each slight deviation, providing that it conferred an advantage in the struggle for existence. This being the case a marked deviation could result from the accumulation, under the influence of natural selection, of numerous small deviations. His conclusion that the numerous slight differences between members of a population—what he called "Individual Differences"—are the raw material of evolution, naturally followed.

After writing the notebooks on the transmutation of species Darwin wrote two essays on the subject, one in 1842 and one in 1844. These were followed by his brief paper to the Linnean Society in 1858 and a year later by his famous book *On the Origin of Species by Means of Natural Selection, or the Preservation of favoured Races in the Struggle for Life*. Each time he set his ideas down on paper his theory of variation became more complex and sophisticated, but by the time he wrote the *Origin* he had chosen Individual Differences for the chief role and had relegated other forms of variation—variation produced by the effects of use and disuse and of hybridisation—to minor roles. Thus in his chapter on "Variation under Nature" he said:

> These individual differences are of the highest importance for us, for they are often inherited, as must be familiar to everyone; and they thus afford materials for natural selection to act on and accumulate, in the same manner as man accumulates in any given direction individual differences in his domesticated productions.[8]

Darwin's next task was to account for these individual differences. Were they, as Pallas maintained, wholly due to the crossing of distinct forms, or as many other authors believed, to the effects of domestication, excess of food, etc.? To answer this question Darwin resorted to his favourite exercise—the collection of as many facts as possible which bear on the question and then to seek, by reasoning from analogy, for a principle or law which unites all the facts. To this end he sought out the breeders. He read their magazines and books, visited their shows and discussed his problems with pigeon fanciers, cage bird enthusiasts and cattle breeders. He came away with a mixture of fact and folklore. All that the breeders could really say about the production of new varieties was that one must use pure breeds, members of different breeds must be crossed and one then selects offspring and repeats this selection of individuals until the new variety is "fixed". Hence Darwin was obliged to admit that:

> When several breeds have once been formed in any country, their occasional intercrossing, with the aid of selection, has, no doubt, largely aided in the formation of new sub-breeds; but the importance of crossing has been much exaggerated, both in regard to animals and to those plants which are propagated by seed.[9]

Now, why was Darwin so loath to attribute much importance to crossing? His answer, given in *The Variation of Animals and Plants under Domestication* contains four reasons: (1) If variability is due to crossing of existing forms one can expect to find fresh combinations of existing characters but no new characters. (2) Such new characters as are supposed to have arisen from crossing may be cases of reversion to long lost characters and therefore not really new. (3) if the various breeds of the domesticated rabbit, for instance, are the modified descendants of a *single* wild species, how can they have arisen from the crossing of *several* distinct forms? (4) Since somatic mutations (he called them "bud sports") arise independently of crossing it is certain that crossing is not necessary for the production of all forms of variation.[10] Darwin also knew that if he admitted Pallas' conclusion he would be destroying his argument for full-scale evolution; since evolution by hybridisation presupposes the existence of the distinct forms which cross, it says nothing about the origin of these distinct forms. To attribute individual differences to the unequal blending of parental characters and to reversions to ancestral characters is only to "push the difficulty further back in time, for what made the parents or their progenitors different?"[11]

We come now to Darwin's reasoning by analogy. In the

Origin he drew a parallel between the effects of changed conditions of life and of crossing. If the change in the conditions of life is a slight one, the effects are beneficial; if it is a radical one the effects are deleterious. As an example of the former he gave the practice of planting the same crop in different fields every year, and of the latter, the sterility of animals when placed under captivity. In an analogous manner, crosses between slightly differing forms enhance the vigour and reproductive capacity of the offspring, but crosses between widely differing forms result in sterile offspring. The reason for this "double-parallelism", as he called it, was clearly that in both cases the reproductive system is disturbed. In 1868 he asked, "Can this parallelism be accidental? Does it not rather indicate some real bond of connection?"[12] In 1876 he gave his answer in *The Effects of Cross and Self-Fertilisation in the Vegetable Kingdom*:

> The most important conclusion at which I have arrived is that the mere act of crossing by itself does no good. The good depends on the individuals which are crossed differing slightly in constitution, owing to their progenitors having been subjected during several generations to slightly different conditions, or to what we call in our ignorance spontaneous variation.[13]

At last he had resolved the variability of hybrids into the effects of the conditions of life and he could justify his utterance of 1868 that ". . . if it were possible to expose all the individuals of a species during many generations to absolutely uniform conditions of life, there would be no variability".[14]

Now it may be objected that the concept of variation which has been outlined here represents Darwin's later more Lamarckian views. A careful study of the *Origin*, however, will not sustain this objection. There Darwin attributed both the individual variability of bisexually reproducing organisms and their sterility "to the same cause",[15] namely—to the reproductive system having been disturbed by changes in the conditions of life. Also in 1859, he drew the double parallel between the good and bad effects of crossing and of changes in the conditions of life.[16] Hence it is clear that as early as 1859 he held the environment responsible for nearly all variation, although he believed that it acts "indirectly", i.e. by way of the reproductive system.

Apart from extracting one general principle—that all variability is in some way connected with changes in the conditions of life—Darwin found the facts of variability difficult to interpret. Nor did he fare any better with inheritance whose laws, he said, "are for the most part unknown".[17] But he argued that since strange and rare deviations such as albinism and the porcupine skin have been shown to be hereditary "less strange and commoner deviations may be freely admitted to be inheritable. Perhaps the best way of viewing the whole subject would be to look at the inheritance of every character whatever as the rule, and non-inheritance as the anomaly".[18] In this way he opened the door to the reacceptance of acquired characters. Indeed, he explicitly challenged the view, which was widely held at that time, that "modifications directly due to physical conditions of life . . . are supposed not to be inherited".[19]

We have seen that Darwin regarded crossing within the species as nature's mechanism for maintaining the uniformity of the species. This was based on the inference of blending heredity. The intermediacy of F1 hybrids was the basis for this inference. Darwin was well aware of non-intermediate hybrids but when he drew up a list of them he found that in most cases "the resemblances seem chiefly confined to characters almost monstrous in their nature, and which have suddenly appeared—such as albinism, melanism, deficiency of tail or horns, or additional fingers and toes; and do not relate to characters which have been slowly acquired through selection".[20]

The dominant theme throughout all Darwin's discussions of variation and heredity was thus the process of the accumulation under selection of small deviations generation by generation. Blending heredity which diminished these deviations at each reproduction was counteracted by the effect of the conditions of life in stimulating fresh variation in each generation. These conditions could be internal—in the womb and in the gynoecium—or external, in the habitat. Darwin did not deny the existence of other forms of variation and heredity; he simply discarded them since they were irrelevant to his argument for evolution.

In the light of these facts it would have been strange indeed if Darwin had arrived at the Mendelian explanation of hybridisation or that he would have appreciated Mendel's point of view had he read his paper. The latter conclusion is supported by the fact that he paid little attention to Charles Naudin's hypothesis of specific segregation in 1863 and to the Mendelian ratios which Galton derived for him in 1875. Naudin is important in the history of genetics not only because he thought of germinal segregation quite independently of Mendel, but also because his work became widely known and was criticised by his contemporaries.

Naudin's Hypothesis of Segregation

Charles Naudin (1815–1899) came to the study of hybridisation as a systematist who sought to use it to clarify the taxonomic relationships between the genera and species of the potato and cucumber families (Solanaceae and Cucurbitaceae). In the 1840s he used hybridisation only for taxonomic purposes, but in the 1850s he took an increasing interest in the evolutionary significance of hybridisation. In 1856 he was struck by the fact that the seedlings from a hybrid *Primula* showed more or less complete reversion to the two species of the original cross. From this empirical observation and from his own *a priori* belief that nature abhors hybrids he hypothesised a process of segregation of the two species within the hybrid. This is what he said:

> Does one not say that Nature is eager to dissolve hybrid forms which do not enter into her plan, and that she does this by the imperfection of the pollen in a large number of hybrids, but also when these hybrids are fertile, by the separation of the two specific essences which art or chance has violently brought together?[21]

Thereupon he sought to demonstrate the complete reversion of the two parental species which should take place if segregation occurs. The majority of these experiments in hybridisation were carried out between 1852 and 1861. In 1860 the Académie des Sciences proposed plant hybridisation as the subject for the *prix des sciences physiques*. Naudin, who by that time had carried out an impressive list of hybridisations, sent in an essay entitled "Nouvelles recherches sur l'hybridité dans les végétaux". The only other entry was that of Godron, Dean of the University of Nancy, but it was Naudin who won the prize. The second part of his essay, containing his conclusions, was published in 1863 in the *Annales des Sciences naturelles*, and the whole essay in 1865 in the *Nouvelles Archives du Musée d'Histoire Naturelle*.

For the experiments described in his prize essay Naudin used 60 different species, 16 belonging to the family Cucurbitaceae, 11 to the genus *Nicotiana*, and 6 to the genus *Datura*. In one case he continued the experiment to the fifth generation and in three cases to the third generation. Though he covered less ground in his experiments than did Koelreuter and Gaertner in theirs, Naudin's work was better in one respect—for each cross he grew many more hybrids. In the preface to the essay Naudin himself remarked on this difference and said that in order to arrive at conclusive results "it is necessary to multiply sufficiently the individuals of the same origin in order to have a chance of encountering all the modifications to which the hybrid forms are susceptible".[22] Evidently he thought conclusive results would follow from the discovery of all the kinds of hybrid variants which are obtainable from a given cross and it did not occur to him that the numbers of these different kinds were related by simple ratios. So he recognised the need for making a representative sample but not for making a rigorous statistical analysis.

Naudin's experimental method was simply to collect seed from a few plants, not taken at random, but selected to show the differing variants of the hybrid generation in question, to sow all the seeds harvested and to grow a convenient number of those which germinated. Unfavourable conditions for germination sometimes reduced his hybrid populations to only a few plants. From fertile hybrids under good conditions he obtained from one to several hundred progeny. His experimental procedure is best illustrated by four examples: *Linaria purpurea* × *L. vulgaris*, *Petunia violacea* × *P. nyctaginiflora*, *Nicotiana rustica* × *N. paniculata* and its reciprocal. For each of these crosses the number of plants grown in each generation and the number from which seeds were gathered are shown in Table I on page 65.

The hybrids whose progeny showed clear reversion were all derived from crosses in the genus *Datura*. This genus has many advantages as genetic material compared with *Nicotiana*. As far as is known none of the species of *Datura* is polyploid, so triploid hybrids are not formed. Several kinds of *Datura* were granted specific status by Linnaeus on the basis of such small differences as flower colour, so that when Naudin made hybrids between such species he was unwittingly carrying out unifactorial crosses. For Naudin working in Paris the genus had one disadvantage. It is half-hardy and therefore, when grown outside, the proportion of seeds which germinate is small

6 *A portion of the letter which Galton wrote to Darwin in 1875 which contains a derivation of Mendelian ratios*

intermediate, viz gray, — thence forward for ever. Then a bit of the tinted structure under the microscope would have a form which might be drawn as in a diagram, as follows:—

white form

black form

whereas in the hybrid, it would be either that some cells were white and others black, & nearly the same proportion of each, thus:—

(1)

giving on the whole when less highly magnified, a uniform gray tint;— or else thus:—

(2)

in which each cell had a uniform gray tint.

7 The Altbrünn Monastery. The 1910 statue of Mendel is on the left-hand side of the picture. The rooms above the arch in the centre of the picture were once occupied by Mendel. The ivy-covered tablet at the far end of Mendel's research plot was erected in 1922 in his honour and is inscribed: 'Praelet Gregor Mendel made his experiments for his law here'.

8a *Gregor Mendel* 1864–65

b *Mendel in* 1862

TABLE I

A selection of Naudin's experimental results

Cross	*Hybrid Generation*	*Population*	*No. of plants from which seeds were harvested*	
Linaria	F1	3	3	
vulgaris ×	F2	400	34	". . . selected from amongst
purpurea	F3	705	122	those which present the
	F4	6	6	most remarkable condi-
	F5	22		tions."
Petunia	F1	1	1	". . . the three plants . . .
nyctaginiflora	F2	47	3	which reproduced best of
× *violacea*	F3	116		all the appearance of the
				variety *albo-rosea*."
Nicotiana	F1	1	1	". . . the nine most con-
paniculata ×	F2	17	9	trasting individuals of the
rusticana	F3	90		second generation."
Nicotiana	F1	36		". . . from those which con-
rustica ×	F2	12	5	trasted to the greatest extent
paniculata	F3	50		with each other."

except in very mild seasons. Also Naudin was unfortunate in encountering a virus which suppresses the formation of spines on the fruits. Thus when he crossed spiny and smooth fruited varieties not all the fruits of the F1 hybrids were covered with spines; some had smooth patches. This led Naudin to make the mistake of extending his process of specific segregation to the body tissues. In 1899 Bateson's *Datura* stock was also infected with this virus and he was unable to account for the resulting "patchy" fruits. The fact that they are due to the "quercina" virus was not established until 1918.

Apart from this shortcoming Naudin's hypothesis was excellent. He saw that if a hybrid forms germ cells of the two parental types there are then three possible combinations between them and these would account for the three types of F2 hybrids: those like the F1 hybrids from the union of dissimilar germ cells, maternal and paternal reversions from like unions of germ cells segregated with respect to the maternal species and with respect to the paternal species. He realised that it was purely a matter of chance as to which pollen grain unites with which ovule, but he did not go on to work out the proportions which one would expect to obtain between the three F2 types if large numbers were produced. On this point and on the question of what is the unit of segregation his hypothesis fell short of Mendel's. Naudin believed that the species segregates as a whole. Consequently his hypothesis can only be applied to crosses involving a single gene difference; for it is only in such "monohybrid" crosses that the F1 hybrids form no more than two types of pollen grains and ovules.

Darwin's Criticism of Naudin's Hypothesis

Naudin and Darwin corresponded from 1862 to 1882, and they exchanged publications. Darwin received the second part of Naudin's prize essay at the end of 1863. He discussed it with Bentham, and then sent Naudin a detailed criticism which is unfortunately not extant today. But his copy of Naudin's essay is preserved at Cambridge. It contains the following marginal comment in Darwin's hand beside the passage on segregation: "This view will not account for distant reversion."[23] By distant reversion he meant the return to characters once apparent but which lay hidden for thousands of generations. The example Darwin usually gave was the one he had himself encountered in his hybridisation experiments; namely, the production of blue feathers and black wingbars, characteristic of the wild pigeon, in brown and white domestic breeds. These wild-type colours had not been known to occur in these fancy breeds as long as they were inbred, and they had been inbred for many years. Crossing these breeds, however, resulted in the appearance of these characters. Darwin attached much significance to distant reversion because it was good evidence for the common descent of all those breeds which reverted to the same "primeval" characters. It was also striking evidence for the almost indefinite persistence of latent characters. This was what prompted him to say in a letter he wrote to Huxley in 1857:

> ... I have lately been inclined to speculate, very crudely and indistinctly, that propagation by true fertilisation will turn out to be a sort of mixture, and not true fusion, of two distinct individuals, as each parent has its parents and ancestors. I can understand on no other view the way in which crossed forms go back to so large an extent to ancestral forms ...[24]

Unfortunately the examples of distant reversion which Darwin cited were really due either to mutation, as in the case of polydactyly, or to genic interaction, as in the case of the wild-type plumage of pigeons. Neither Naudin's nor Mendel's theory could be used to explain these cases. And when Mendel's theory was restated in 1900 by Bateson, his opponents, the biometricians, were quick to point out that it did not account for the reappearance of wild-type agouti coat-colour in the offspring from the crossing of albino and piebald breeds of mice. Hence it seems unlikely that Darwin would have been any more favourably disposed towards Mendel's theory, had he known of it, than were his very Darwinian successors the biometricians.

The sheer weight of empirical evidence prevented Darwin from ruling out non-blending heredity altogether and caused him to incorporate Naudin's hypothesis of segregation into his theory of heredity in order to account for the behaviour of hybrids; but he continued to think in terms of the blending theory. The assumptions of blending heredity were: (1) each parent contributes equally to the offspring, (2) these contributions are halved at each successive generation. This repeated halving is represented in Table II which is extracted from a similar table in Charles White's book *The Regular Gradation of Man* . . . 1799.

TABLE II

Terminology of the members of mixed races and their genetic constitution in terms of "blood" fractions

Parents	*Offspring*	*Genetic Designation*	*Degree of Mixture*
Negro and European	Mulatto	F1	$\frac{1}{2}$ white $\frac{1}{2}$ black
European and Mulatto	Terceron	1st back-cross	$\frac{3}{4}$ „ $\frac{1}{4}$ „
European and Terceron	Quarteron	2nd „ „	$\frac{7}{8}$ „ $\frac{1}{8}$ „
European and Quarteron	Quinteron	3rd „ „	$\frac{15}{16}$ „ $\frac{1}{16}$ „

According to R. C. Punnett the representation of blending heredity as a series of fractions was adumbrated as early as 1722 by William Wollaston (1660–1724), but it was Francis Galton (1822–1911) who gave it the precise formulation of: Heritage $= \frac{1}{4}$ p $+ \frac{1}{8}$ pp $+ \frac{1}{16}$ ppp . . . (where p = parent, pp = grandparent and so on) and who called it the Ancestral Law of Inheritance. This law could account for distant reversion more readily than could the law of segregation since it assumes that no ancestral contribution is ever lost, it is only diminished. On the theory of segregation two grandparental contributions (i.e. one member of each pair of chromosomes) are always lost at each reproduction, one from each germ cell. To account for distant reversion on the Ancestral Theory one has to postulate that even a minute fraction of a distant ancestral contribution can produce a sensible effect. Having regard to the powerful effects of infinitesimal quantities of certain chemicals on living organisms Darwin did not find this postulate absurd. On the Mendelian theory one has to postulate either that the wild-type character is only expressed when two quite independent genes are present, say *A* and *B*, and that in one domestic breed *A* has mutated to *a* and , in the other, *B* has mutated to *b*. Then it is not until the two breeds *aaBB* and *AAbb* are crossed that *A* and *B* are brought together again to give the wild-type combination of genes. Or, in cases such as polydactyly, where it is not the crossing of different breeds which results in the reappearance of long-lost characters, we believe that it is due to mutation. If one of the genes in man has a tendency to mutate in the same manner from time to time then one would always expect to find a very few polydactyle individuals in a large population; and this is so.

We have seen that Darwin attributed all variation ultimately to the conditions of life which he believed act both on the reproductive system and on the individual in its embryonic condition and even at later stages in its development. The results of this action were, he assumed, inherited. Therefore it became necessary to suggest a mechanism for the inheritance of these acquired characters, so he postulated the existence of genetic particles which he called "gemmules". These are generated by the body tissues and are sent to the reproductive organs via the circulating fluids. His cousin, Francis Galton, put Darwin's assumption of the free circulation of the gemmules to the test of experiment by inter-transfusing the blood of rabbits whose coats were differently coloured, and thus mixing the gemmules. His aim was to see whether the resulting "polluted" rabbits when inbred produced offspring whose coat colour was tainted. They did not, so Galton was not inclined to accept Darwin's theory.

Galton's Approach to Mendelian Heredity

Galton's views on inheritance were far more modern than those of Darwin. He gave a lecture on the subject in 1875 to the Anthropological Institute. It was subsequently published in the Institute Journal under the title of "A Theory of Heredity". Galton sent his cousin a copy in November. Here he expressed very clearly the doctrine of the continuity of the germplasm which he called "stirp" (L. *stirps* = stock, stem, hence a line of descent). Darwin found the paper difficult to understand but his son George was able to make clear to him the ways in which the theory there expressed differed from pangenesis. The chief difference was over the source of hereditary units. Galton maintained that they multiply in the reproductive organs and that they receive very little if any units from the rest of the body tissues. Darwin held that all the tissues contribute. On December 18th, 1875, he wrote to Galton asking him how, on his view, he would explain the following fact:

> If two plants are crossed, it often or rather generally happens that every part of [the] stem, leaf—even the hairs—and flowers of the hybrid are intermediate in character; and this hybrid will produce by buds millions on millions of other buds all exactly reproducing the intermediate character. I cannot doubt that every unit of the hybrid is hybridised and sends forth hybridised gemmules. Here we have nothing to do with the reproductive organs. . . .[25]

Darwin was interested in the asexual propagation of hybrids because, as he said in the *Origin*, "With plants which are temporarily propagated by cuttings, buds, etc., the importance of crossing [for evolution] is immense;

for the cultivator may here disregard the extreme variability both of hybrids and of mongrels . . ."[26]

Galton replied as follows:[27]

Dec. 19/75.

My Dear Darwin,

The explanation of what you propose, does not seem to me in any way different on my theory, to what it would be on any theory of organic units. It would be this:—

Let us deal with a single quality, for clearness of explanation, and suppose that in some particular plant or animal and in some particular structure, the hybrid between white and black forms was exactly intermediate, viz. gray,—thenceforward for ever. Then a bit of the tinted structure under the microscope would have a form which might be drawn as in a diagram, as follows:—(see Plate 1) whereas in the hybrid it would be either that some cells were white and others black, and nearly the same proportion of each, thus:—(see (1), Plate 2) giving *on the whole* when less highly magnified a uniform gray tint,—or else, thus:—(see (2), Plate 2) in which each cell had a uniform gray tint.

In (1) we see that each cell had been an organic unit (quoad colour). In other words, the structural unit is identical with the organic unit.

In (2) the structural unit would not be an organic unit but would be an organic *molecule*. It would have been due to the development, not of one gemmule but of a group of gemmules, in which the black and white species would, on statistical grounds, be equally numerous (as by hypothesis they were equipotent).

The larger the number of gemmules in each organic molecule, the more *uniform* will the tint of grayish be in the different units of structure. It has been an old idea of mine, not yet discarded and not yet worked out, that the number of units in each molecule may admit of being discovered by noting the relative number of cases of each grade of deviation from the mean grayness. If there were two gemmules only, each of which might be white or black, then in a large number of cases one quarter would be always quite white, one quarter quite black, and one half would be gray. If there were 3 molecules, we should have 4 grades of colour (1 quite white, 3 light gray, 3 dark gray, 1 quite black and so on according to the successive lines of "Pascal's triangle"). This way of looking at the matter would perhaps show (a) whether the number in each given species of molecule was constant, and (b), if so, what those numbers were.

Ever very faithfully yours,

Francis Galton.

Here we find all the elements of the Mendelian explanation save the independent segregation of different pairs of characters. The hereditary units brought together in the hybrid have not fused. There are a definite number of them. For any character difference the number can be determined by finding the "relative number of cases in each grade . . ."—Mendel's ratios. Inheritance may be simple, two units per character, giving the 1 : 2 : 1 ratio (where dominance occurs, as in Mendel's experiments with *Pisum*, the apparent ratio is 3 : 1); or it may be more complex as in Galton's next ratio 1 : 3 : 3 : 1.

When we compare this letter of Galton's with Mendel's paper on *Pisum* a difference of approach is apparent. The argument is the same but the starting-point is not. Mendel starts with the 3 : 1 ratio derived from precise experiments involving known crosses and he goes on to infer the presence of two different hereditary units which combine in all possible ways an equal number of times. Galton starts with the concept of a finite number of hereditary units and then describes how one might be able to determine the number and degree of constancy of these units from observational data. But when he speaks of noting the number of "cases" of each grade is he referring to an outbreeding population in nature or to the offspring from the crossing of white and grey individuals? Presumably he refers to the latter since it is the subject of the first part of the letter. Really the difference is not so great as it

TABLE III

Pascal's Triangle

1

1 1

1 2 1

1 3 3 1

1 4 6 4 1

1 5 10 10 5 1

1 6 15 20 15 6 1

1 7 21 35 35 21 7 1

1 8 28 56 70 56 28 8 1

appears at first. Mendel wrote his paper on *Pisum* at a time when it was just not done to present a scientific theory in a non-inductive manner. The theory must flow from the experiments and observations; but the fact that Mendel started by looking for statistical relationships in the F2 generation suggests that he had worked out a hypothesis on the lines of Galton's letter. Like the latter, Mendel was happy to write in mathematical language and to think in abstract terms. Consider this passage for instance:

> If n represents the number of differentiating characters in the two original stocks, 3^n gives the number of terms of the combination series, 4^n the number of individuals which belong to the series, and 2^n the number of unions which remain constant. The series therefore contains, if the original stocks differ in four characters, $3^4 = 81$ classes, $4^4 = 256$ individuals, and $2^4 = 16$ constant forms . . .[28]

Mendel could have arrived at the 1 : 2 : 1 ratio as Pascal did in the seventeenth century from working out the probable relative frequencies of the various combinations of two entities. They are given by the coefficients of the binomial $(a+b)^2 = a^2 + 2ab + b^2$. Pascal fitted the coefficients of successively higher binomials into the triangle which Galton referred to in his letter. It is shown in Table III. The third row from the top gives the coefficients of $(a+b)^2$, and covers all cases involving two entities which can be combined in three different ways.

The fifth line of the triangle contains the figures 1, 4, 6, 4, 1. These are the ratios which one would obtain if one character difference were determined by four elements all of equal potency. Thus to take Galton's example, if a cross is made between a black organism possessing the determinants BBBB and a white organism possessing the determinants AAAA then the hybrid will have the hereditary constitution AABB and the familiar Punnett Diagram setting out the number of different possible combinations which can occur when the hybrid reproduced is shown in Table IV on page 75.

From it we get the ratio: 1 black : 4 dark grey : 6 grey : 4 light grey : 1 white. These frequencies add up to 16 just as do the frequencies for a di-hybrid cross showing dominance—9 : 3 : 3 : 1—which Mendel demonstrated experimentally. Mendel's ratio is only a modification of the frequencies given in the fifth line of Pascal's triangle.

Galton saw what the consequences would be of random

TABLE IV

Punnett Diagram illustrating the classes of F2 progeny resulting from the hypothetical cross: black × white, in which segregation is in accordance with the fifth line of Pascal's Triangle

Gametes	AA	AB	AB	BB
AA	AAAA	AAAB	AAAB	AABB
AB	AAAB	AABB	AABB	ABBB
BA	AAAB	AABB	AABB	ABBB
BB	AABB	ABBB	ABBB	BBBB

fertilisations between different germ cells and yet he did not attempt to demonstrate it. Why didn't he? Was this letter to Darwin just a "flash in the pan", suddenly arrived at and soon forgotten? It was certainly not. In 1872 he had published a wonderfully perceptive paper which he described to Darwin very modestly as "a little paper to be shortly read at the Royal Society on Blood-relationship in which I try to define what the kinship really is, between parents and their offspring." Here he made such prophetic statements as:

> . . . each individual may properly be conceived as consisting of two parts, one of which is latent and only known to us by its effects on his posterity, while the other is patent, and constitutes the person manifest to our senses.
>
> The span of the true hereditary link connects, as I have already insisted upon, not the parent with the offspring, but the primary elements of the two, such

> as they existed in the newly impregnated ova, whence they were respectively developed.
> ... we gratuitously add confusion to our ignorance, by dealing with hereditary facts on the plan of ordinary pedigrees—namely, from the persons of the parents to those of their offspring.
> ... It is often remarked (1) that the immediate offspring of different races or even varieties resemble their parents equally, but (2) that great diversities appear in the next and succeeding generations ... A white parent necessarily contributes white elements to the structureless stage of his offspring and a black, black; but it does not in the least follow that the contributions from a true mulatto must be truly mulatto.[29]

Galton's derivation of the 1 : 2 : 1 ratio for a cross between white and black organisms in his letter of 1875 may be looked upon as the sequel to this statement of 1872 on mulattos. Hence there is no doubt that Galton had worked out the Mendelian explanation of hybridisation and that he regarded it as a possible explanation although he never sought to demonstrate it. Possibly he first thought of it after reading Naudin's paper on plant hybridisation which Darwin sent him in 1870. His reason for not repeating any of Naudin's experiments was, he said, because he was "too ignorant of gardening, and living in London with a summer tour in prospect I don't see my way to a successful issue . . ."[30]

After 1875 Galton's genetical studies were chiefly devoted to formulating a mathematical theory of inheritance. In his book *Hereditary Genius* of 1869 he had shown that for characters such as intelligence the mean of a population remains the same in successive generations—there is, to use Galton's phrase, a "stability of type". The distribution of deviations from the mean follows "the very curious theoretical law of 'deviation from an average' " which, "M. Quetelet, the Astronomer Royal of

Belgium, and the greatest authority on vital and social statistics, has largely used in his inquiries".[31] Now Galton believed that from a study of the distribution of deviations in successive generations it should be possible to deduces the laws of inheritance.

In order to find these laws he carried out an experiment from which he wanted "suggestion", not proof, "because the theoretical exigencies of the problem would afford that".[32] He decided to study the inheritance of seed size. He used sweet peas because he thought they were self-fertilising. He sorted 630 seeds, taken from one seedman's bin and presumably all belonging to the same variety, into seven grades according to their diameter. To nine co-operative friends he sent ten seeds of each grade. They sowed the seeds of each grade separately, harvested their pods separately and returned them to Galton.

Galton compared the variation in the progeny from seeds of each parental grade. The mean diameter of the progeny from large seeds was smaller than that of their parents. The mean diameter for all the progeny was less than the parental mean. There had been a reversion or "regression" as he called it, towards mediocrity or the "racial mean".

Galton's experimental technique can be criticised on many points. Certainly it was very inferior to the technique of Mendel. Yet Galton complained afterwards that he had taken immense pains over this experiment and he doubted that he would have taken as much trouble if he had understood the general conditions of the problem then as clearly as he did later. Compared with the trouble Mendel took over his experiments Galton's "immense pains" seem trifling. Thus, if Galton's plants yielded 80 seeds each he would have had to measure nearly 40,000 seeds. (Since two crops failed he was able to consider only the produce from 630 parental seeds.) By measuring the seeds in tens he reduced the number of operations ten times, yet Mendel, with his unifactorial crosses alone, examined 15,347 seeds. In addition, Galton did not grow his own plants.

When he interpreted the results of this experiment he assumed that the deviations he recorded were subject to heredity. Then he had to explain why large seeds did not have equally large offspring, so he postulated a tendency which opposed the expression of heritable deviations. In this way he believed he had explained why successive generations maintained statistical identity. Really he had only shown either that his original sample was heterozygous or that the deviations he measured were not inherited.

Galton did not even consider these alternative explanations but proceeded to calculate the value of regression in the case of the sweet pea. He argued that since these plants were self-fertilising the value of one-third which he obtained was equal to the regression of offspring on a single parent. When he calculated the regression in stature of man he expected an answer of two-thirds—the value for the regression of offspring on two parents. In fact sweet peas are regularly cross-pollinated and regression on 2 parents is not twice but $\sqrt{2}$ times the regression on a single parent. Karl Pearson has shown how Galton obtained this value of 2/3 and has justly criticised the method and corrections which Galton employed. Still more unconvincing is Galton's attempt to derive the fractional theory of heredity from his regression coefficients. As early as 1865 he had assumed this theory to be correct and when he derived it from his regression coefficients he was satisfied that it merited the title "Ancestral Law of Inheritance".

Thus in the twilight period before the rediscovery of Mendel's laws Galton had clothed blending heredity with a precise formulation and brought it into the centre of the scientific arena. His evidence for the ancestral influences asserted by his ancestral law was simply the stability of types. If these influences did not exist it would be possible by repeated selection of tall men to breed giants—"the giants (in any mental or physical particular) would become more gigantic, and the dwarfs more dwarfish, in each successive generation".[33]

We can see now that Galton's error had been to attribute too much significance to his regression and correlation coefficients. After all, a correlation coefficient of +1 between two sets of measurements tells us merely that deviations in one set of measurements are exactly paralleled by deviations in another. It may tell us, for example, that tall men have long femurs and short men have short femurs. It does not demonstrate the existence of a causal relationship between height and length of femur. Still less does a coefficient of regression of one-third between parent and offspring tell us that the smaller average deviation of the offspring compared with that of their parents is due to the weight of ancestral contributions which "dilutes" the influence of the parental deviation. But Galton thought this was the case. He went further and asserted that the halving of ancestral contributions is brought about by the reduction division which precedes the formation of the germ cells.[34]

August Weismann (1834–1914), the Professor of Zoology at Freiburg was content, like Galton, to accept the fractional law of inheritance. In his essay "The Continuity of the Germ-plasm as the foundation of a theory of Heredity", 1885, he calculated the proportional contribution of each generation

> . . . after the manner in which breeders, when crossing races, determine the proportion of pure blood which is contained in any of the descendants. Thus while the germ-plasm of the father or mother constitutes half the nucleus of any fertilised ovum, that of a grandparent only forms a quarter, and that of the tenth generation backwards only 1/1024, and so on. The latter can, nevertheless, exercise influence over the development

> of the offspring, for the phenomena of atavism show that the germ-plasm of very remote ancestors can occasionally make itself felt.[35]

Consequently Weismann had to find a *transverse* form of division such as would split in half the parental germ plasm situated along the chromosomes. At the same time Flemming's discovery of the longitudinal division of the chromosomes in mitosis had to be accepted for the evidence was compelling, so Weismann had a strong *a priori* reason for finding a transverse division in meiosis, and here van Beneden and Carnoy came to his aid. They saw what looked like a transverse division in meiosis I. In fact they had only misinterpreted the striking falling apart and contraction of homologous chromosomes which precedes their migration to opposite poles of the nuclear spindle, and is so much more marked a feature of meiosis than of mitosis.

Flemming's discovery of the longitudinal division of chromosomes in mitosis led Weismann to conclude that the hereditary units or "ancestral germ-plasms" are "upon the whole, arranged in a linear manner in the thin thread-like loops" [i.e. chromosomes], and he regarded the "longitudinal splitting" of these loops as "almost a proof of the existence of such an arrangement, for without this supposition the process would cease to have any meaning".[36] But as we have seen, he went on to mar this accurate prophecy by insisting on the existence of transverse division as well. He was led astray by the theory of the breeders and the observations of van Beneden and Carnoy. We need not therefore be surprised that Galton was misled in the same way.

Galton's genetical studies should not be judged simply by comparison with Mendel's, for Galton was above all concerned with the question of how evolution can take place in a population. He deserves credit for realising that on the evidence at his disposal it is "impossible that the

9 *Mendel surrounded by members of the excursion party from Vienna to the 1862 London Industrial Exhibition. On their way to London they stopped in Paris where a photograph of the entire party was taken in front of the Grand Hotel, on 6 August*

Geum intermedium:
ABcDEe
ABcdEe

Geum urbanum

G. rivale

pag 307

10 *Mendel's notes on* Geum urbanum, rivale *and* intermedium *at the back of his copy of Gaertner's book. (See pp. 185–186)*

11 *Mendel's notes on the results of an experiment which did not fit his theory.* (*See* p. 186)

~~V1 + gV1 = 112 Violett~~ 7?

V1 = 37
g = 37
gV1 = 75
V1W = 150
gW = 150
W = 150

V1W + gW = 300 weiß Viol 250 −50
W = 150 Weiss 166 +16
gV1 = 75 6 B 65 −10
g = 37 d B 27 −10
V = 37 Viol 93 +56

343 6V&V 351 7/12
92 B 100 1/6 2/12
66 W 150 1/4 3/12

Woll
Loewe t
~~traum~~
etwas
Lehre l
wer auch die Welt will zu
Der sich hübsch bü

: 305 = 59 : 296
340
179 95 : 296 = 61

75
150
1/4 [illegible] 6

W 150 1/4 W
bB 75 1/8 gV1
dB 37 1/16 g
6V 300 1/2 gW + V1W
V 37 1/16 V.

Pisum arvense 1blüthig, die Flügel roth
P. arvense & sativum Hülsen fast walzig, walzig u. gerade
bei P. umbellatum Mill.; gerade schwertförmig eingedrückt beiderseits bei saccharatum Host. (var. flexuosum
Willd. sichelförmig [illegible] u. [illegible], Samen klein,
eckig); gerade schwertförmig, nicht eingedrückt,
Samen dicht an einander gepreßt bei P. quadratum Mill.
bei P. sativ. & arvense Nebenblätter unten abgerundet
und fein gezähnt-gekerbt, halbherzförmig.
bei saccharat. & quadrat. Nebenblätter schief einwärts
u. rundgeschnitten, Hülsen breit zusammengedrückt
bei sativ. ~~quad~~ saccharat. & umbellat. Samen kugelig

12 *Mendel's notes on* Pisum *at the back of his copy of Gaertner's book.* (*See p.* 187)

natural qualities of a race may be permanently changed through the action of selection upon mere variations".[37] It is also to his credit that he recognised a distinction between continuous variations which he called "variation proper" or "mere variations" and discontinuous variations which he called "sports". Only the latter, he held, constitute the raw material for evolution, since they involve "a change of typical centre, a new point of departure has come into existence, towards which regression has henceforth to be measured, and consequently a real step forward has been made in the course of evolution".[38]

In 1894 Bateson's book *Materials for the Study of Variation* appeared. In it Bateson coined the terms "Continuous" and "Discontinuous" variation[39] and asserted that evolution can only take place by the latter. Galton read it with the "utmost pleasure" and complained that his own very similar views which he had put forward as early as 1869 had met with no response. "I seemed to have spoken to empty air," he said. "I never heard nor have I read any criticism of them, and I believed they had passed unheeded and that my opinion was in a minority of one."[40]

Galton was probably the most original biologist of the century but he was by nature shy, so he did not push an idea if it failed to excite the interest of his contemporaries, especially that of his cousin Darwin. The Ancestral Law, on the other hand, found a welcome reception and in all his investigations of heredity whether of characters which blend (height in man) or partially blend (eye colour in man), or refuse to blend (chestnut and bay horses and lemon-white and tri-colour Basset Hounds) he sought to demonstrate the Ancestral Law from the data collected.

Today Galton's Ancestral Law of Inheritance still stands as a mathematical representation of the average distribution of continuously varying characters in a population of freely outbreeding individuals not subject to selection. It serves as a basis for predicting the average distribution of such characters in the population. It tells us that on an average a grand-parental deviation will be diminished to one-eighth of its original magnitude in the grandchildren. The Mendelian theory, on the other hand, tells us that only one in eight grandchildren will have received this grandparent's genes for the said deviation. Expressed as averages for a population, however, both theories give the same prediction.

The blending theory as a theory of heredity belongs to the history of error, but it is nonetheless important in the history of biology. It caused Darwin to postulate an unnecessarily high mutation rate and it side-tracked Galton from the study of non-blending or "alternative" inheritance. Finally in 1900 it was blending heredity and Galton's formulation of the Ancestral Law which the biometricians backed to the hilt in their fight against Mendelism. From the evolutionary standpoint, characters which blend were the ones to study, but Darwin would have done well to heed Galton's warning that in the lines of evolutionary descent "the changes are not by insensible gradations; there are many, but not an infinite number of intermediate links",[41] and Huxley's criticism that in excluding saltatory evolution "you have loaded yourself with unnecessary difficulty".[42]

The principle of evolution by many small but definite steps which Galton stated in 1869 is accepted today as the chief form of evolution. Why did Darwin not accept it? I think he was biased against saltatory evolution in any form because the majority of its supporters—Lyell and Harvey for instance—were thinking in terms of mutations large enough to produce the discontinuities between species. Therefore they denied the role of selection in accumulating the many intermediate steps which on Darwin's view are necessary if adaptive divergence is to result. To the mutationists these intermediate steps just did not exist. And the mutationists at the beginning of the present century belittled natural selection just as Darwin had feared.

Thus we see that it was the need to account for adaptive variation in a mechanistic manner which prevented Darwin from espousing any form of saltatory evolution. Behind this attitude we can discern the geologist and palaeontologist in Darwin dictating the law *Natura non facit saltum*. So strong was his feeling on this point that he prophesied in 1859 that just as ". . . modern geology has almost banished such views as the excavation of a great valley by a single diluvial wave, so will natural selection, if it be a true principle, banish the belief of the continued creation of new organic beings, or of any great and sudden modification in their structure."[43]

NOTES

(1) Barlow, N. 1958. *The autobiography of Charles Darwin 1809–1882 with original omissions restored. Edited with Appendix and notes by his grand-daughter Nora Barlow*. London. p. 77.

(2) Ibid., p. 128.

(3) Knight, T. A. 1799. "An account of some experiments of the fecundation of vegetables." Phil. Trans., *89*, 202.

(4) Sprengel, C. K. 1793. *Das entdeckte Geheimniss der Natur im Bau und in der Befruchtung der Blumen*. Berlin. p. 43.

(5) De Beer, Sir G. (Ed.) 1960. "Darwin's notebooks on the transmutation of species part IV." Bull. Br. Mus. nat. Hist. F. Historical, *2*, No. 5, 164. (Darwin's notebooks on the transmutation of species were published in the above periodical, vol. ii, Nos. 2–5, and are referred to in future as N1, N2, N3, N4.)

(6) Darwin, C. R. 1872. *The origin of species by means of natural selection or the preservation of favoured races in the struggle for life*. 6th ed., reprinted in the Everyman Library under the title *The Origin of Species*. 1958, p. 58. (Referred to in future as O6e.)

(7) O6e, p. 51.

(8) O6e, p. 51. (Quotations (6) and (7) do not appear in the 1859 edition. Quotation (8) is a slightly expanded version of the passage on p. 39 of the 1st edition: *On the origin . . . reprinted* with a foreword by Prof. C. D. Darlington. Watts, London. 1950. Referred to in future as O1e.)

(9) O6e, pp. 48–49. O1e, p. 37 (almost identical).

(10) O6e, p. 24. O1e, p. 8.

(11) Darwin, C. R. 1868. *The variations of animals and plants under domestication*. 2 vols. London. Vol. ii, p. 252. (Referred to in future as VuD.)

(12) VuD, vol. ii, p. 177.
(13) Darwin, C. R. 1876. *The effects of cross and self fertilisation in the vegetable kingdom*. London. p. 27.
(14) VuD, vol. ii, p. 255.
(15) O1e, p. 8.
(16) O1e, pp. 226–228.
(17) O6e, p. 26. O1e, p. 11.
(18) O6e, p. 26. O1e, p. 11. (Almost identical.)
(19) O6e, p. 50. O1e, p. 39.
(20) O6e, p. 288. O1e, p. 253. (Almost identical.)
(21) Naudin, C. 1856. "Observations constatant le retour simultané de la descendance d'une plante hybride aux types paternels et maternels." C. R. Acad. Sci., Paris, *42*, 628.
(22) Naudin, C. 1865. "Nouvelles recherches sur l'hybridité dans les végétaux." Arch. Mus. Hist. nat., Paris, *1*, 26.
(23) Naudin, C. 1863. "Nouvelles recherches sur l'hybridité dans les végétaux." Ann. Sci. Nat. Botanique, sér. 4, *19*, 193. Annotation by Darwin—Cambridge Reprint No. 300.
(24) MLD, vol. i, p. 103.
(25) MLD, vol. i, p. 362.
(26) O6e, p. 49. O1e, p. 37. (Almost identical.)
(27) Pearson, K. 1914–1930. *The life letters and labours of Francis Galton*. 3 vols. Cambridge. Vol. ii, pp. 189–190. (Referred to in future as LFG.)
(28) MR1, p. 19.
(29) Galton, F. 1872. "On blood-relationships." Proc. roy. Soc., *20*, 394 (1st quotation), 400 (2nd quotation), 401 (3rd quotation), 402 (4th quotation).
(30) LFG, vol. ii, p. 159.
(31) Galton, F. 1892. *Hereditary Genius. An inquiry into its laws and consequences.* With an introduction by C. D. Darlington. Fontana Reprint, 1962. London. pp. 66–67. (Referred to in future as HG.)
(32) Galton, F. 1877. "Typical laws of inheritance." Nature, *15*, 512.
(33) HG, p. 33.
(34) Galton, F. 1897. "The average contribution of each several ancestor to the total heritage of the offspring." Proc. roy. Soc., *61*, 403.
(35) Weismann, A. 1885. "The continuity of the germplasm as the foundation of a theory of heredity." In: *Essays upon heredity and kindred biological problems by Dr. August Weismann.* Trans. Poulton, Schönland & Shipley. Oxford. 1889. p. 179.
(36) Weismann, A. 1887. "On the number of polar bodies and their significance in heredity." In: *Essays upon heredity . . .*, pp. 359–60.
(37) HG, p. 34.
(38) HG, p. 34.
(39) Bateson, W. 1894. *Materials for the study of variation treated with especial regard to discontinuity in the origin of species*. London. p. 15.
(40) Galton, F. 1894. "Discontinuity in evolution." Mind. N.S., *3*, 369. (Review of Bateson's book.)
(41) HG, p. 421. (This passage comes in the final section entitled "General Considerations" which is also in the original edition of 1869.)

(42) Huxley, L. 1900. *Life and letters of Thomas Henry Huxley.* 2 vols. London. Vol. 1, p. 254.
(43) O1e, p. 82. O6e, p. 94.

SUGGESTIONS FOR FURTHER READING

Blaringham, L. 1913. "La notion d'espèce et la disjonction des hybrides d'après Charles Naudin (1852–1875)." Progressus Rei Botanicae, *4*, 27–108.

Darlington, C. D. 1959. *Darwin's place in history.* Basil Blackwell, Oxford; 1961, Macmillan, New York.

Ehrlich, P. R. and Holm, R. W. 1963. *The process of evolution.* McGraw-Hill, New York and London.

Eiseley, L. 1959. *Darwin's century. Evolution and the men who discovered it.* Victor Gollancz, London; 1958, Doubleday, New York.

Fisher, R. A. 1958. *The genetical theory of natural selection.* 2nd ed. Dover Publications, New York.

Sheppard, P. M. 1958. *Natural selection and heredity.* Hutchinson, London; Hillary, New York.

Vorzimmer, P. 1963. "Charles Darwin and blending heredity." Isis, *54*, pt. 3, No. 177, 371–390.

Chapter Four

Sexual and Asexual Reproduction

At the beginning of the nineteenth century it was known that reproduction can take the form of the division of parts, or budding, which is purely vegetative and is therefore termed asexual, and the form of eggs and ovules which involves the sexual act and is therefore termed sexual. At that time the distinction was quite clear, being based on the presence or absence of fertilisation and of variation. The property of bisexual reproduction in yielding new varieties had been exploited by the early animal breeders in the eighteenth century, notably by Bakewell, and a little later by the plant breeders led by Thomas Andrew Knight. Thus we find mention of this distinctive role of bisexual reproduction in the writings of Charles Darwin's grandfather, Dr. Erasmus Darwin. In his *Zoonomia* . . . of 1794 he states that offspring in the form of buds and bulbs

> . . . exactly resemble their parents, as is observable in grafting fruit trees, and in propagating flower-roots; whereas the seminal offspring of plants, being supplied with nutriment by the mother, is liable to perpetual variation.[1]

It was also known that seminal hybrids are often intermediate between the originating species. This fact suggested that an intimate fusion of the two forms occurs.

Graft hybrids, on the other hand, consist of the two forms which grow side by side but remain distinct. This fusion appeared therefore to be unique to sexually produced hybrids, thus giving a third distinction between the two forms of reproduction.

Now why is the distinction important? Why not look upon all forms of reproduction as fundamentally the same, if you like, as various forms of growth? The answer is that sexual reproduction is a unique process, bisexual reproduction is the chief source of variation and the only means of modifying and distributing mutations within a population. Hence it is essential that the evolutionist should recognise its distinctive nature. Darwin refused to do this for three reasons. His first reason was an *a priori* one—his theory of variability required that the distinctive nature of sexual reproduction be denied. This has been discussed in Chapter 3. His second reason was that there are many exceptions to the established distinctions. In this chapter we will describe these exceptions and then see how Darwin's notorious hypothesis of Pangenesis arose from his third reason—a desire to account for both the distinctions and the exceptions—to harmonise the conflicting evidence.

The necessity of fertilisation

In the eighteenth century Charles Bonnet had discovered that eggs belonging to the spring and summer broods of the greenfly or aphid require no fertilisation before developing. Henceforth this process of virgin birth was termed parthenogenesis. In 1845 the Silesian bee-keeper, Pastor Dzierzon, put forward his hypothesis of the formation of drone honey bees from unfertilised eggs of the queen.[2] In the 1850s he established it on the basis of hybridisation experiments between German and Ligurian bees.[3] In 1856 von Siebold published his book *On a true Parthenogenesis in moths and bees*. He began by conducting a critical examination of the cases of parthenogenesis so far reported in order to find out "whether we have to do here with credible facts, or whether, in this case, a fact has not been rather concluded from superficial, unsatisfactory and scanty observations, than positively proved".[4] From this analysis he eliminated several supposed cases but affirmed the case of the honey bee and those of four moths. This result, he admitted, was not what he had expected. Consequently the "admitted proposition of the fecundation theory, that the development of the eggs can only take place under the influence of the male semen", he said, "has suffered an unexpected blow by Parthenogenesis".[5] Darwin, who corresponded with von Siebold, regarded his opinion as representative of that of most physiologists. Hence we see why he began to doubt this distinction.

Nor was parthenogenesis confined to animals. John Smith, who was curator of the Royal Botanic Gardens, Kew, from 1841 to 1864, read a paper to the Linnean Society in 1839 entitled "Notice of a plant which produced perfect seeds without any apparent action of pollen." Three specimens of this plant, *Alchornia ilicifolia*, popularly known as "Dovewood", had been brought to Kew from Australia, its native home, in 1829. For several seasons they produced only female flowers and yet they yielded fruits and seeds. Smith germinated the seeds and exhibited the resulting plants to members of the Linnean Society. These young plants were so like the parents as to preclude any likelihood that they had been pollinated by other species in the vicinity of the three mother plants and there were no male plants in England at the time. Moreover, John Smith could find no pollen tubes in the styles of the female flowers. After considering these facts, he said, "I can arrive . . . at no other conclusion than that pollen is not essential to the perfecting of its seeds".[6] Other cases of such apomictic species were only discovered slowly so that by the end of the nineteenth century the list comprised but a handful of species. The case which proved the most notorious was that of *Hieracium*. Naegeli and Mendel succeeded in raising hybrids in this genus which bred true, a fact which Mendel was unable to explain on his theory. Little did he know that his hybrids were reproducing parthenogenetically and therefore segregation did not occur.

Some animals and plants can produce in an apparently sexual manner but no fusion of gametes actually takes place. The egg requires the stimulus of the sperm but not its material contribution, a condition termed "pseudogamy". This frequently happens in the genera *Rubus* and *Rosa* when interspecific crosses are attempted. The offspring, of course, are all identical with the mother plant.

The first report of pseudogamy was provided by Marie A. Millardet (1838–1902), a professor in Bordeaux. In 1894 he published his now famous paper "Note sur l'hybridation sans croisement ou fausse hybridation", which contains his account of the crossing of various varieties of strawberries made between 1883 and 1893. The majority of the offspring were identical with the mother plants, thus it seemed that hybridisation had not actually occurred. Millardet believed it had, but that the male element was completely suppressed. As further examples of the same state of affairs he cited the 'biased' tobacco hybrids which Gaertner obtained (see Chapter 2) and the completely dominant F1 hybrids which Naudin obtained when he crossed certain thorn apple species (see Chapter 3). He saw nothing impossible about these facts despite their strangeness. To him they represented the "extreme term of a series of perfectly established facts".[7] On the one side one has intermediate hybrids, then nearly intermediate hybrids, next biased hybrids and finally his strawberry hybrids which he termed false hybrids ("faux hybrides").

Millardet was, of course, unaware that the contents of the pollen tube had failed to fuse with the egg nucleus. As a result he combined and confused the quite distinct phenomena of genetic dominance, and pseudogamy, much in the same way as Darwin did in the case of quite distinct forms of variation.

On the basis of our knowledge of the mechanics of chromosomes a series can be constructed which in a sense bridges the gap between buds and eggs. On the one side we have the meristematic cell of a vegetative bud, on the other, the zygote resulting from the fusion of two reduced germ cells. In between these two extremes we have the reduced pseudogamous egg cell, the reduced parthenogenetic egg cell, and the unreduced parthenogenetic egg cell which arises either as a result of an incomplete meiosis followed by reconstitution or by the complete suppression of meiosis. The latter is distinguished from the meristematic cell of a bud by a number of cytoplasmic features but the nuclei of the two cells are identical.

The Presence of Variation

Gardeners noted that from time to time buds gave rise to new varieties. These bud varieties or "bud sports" are well known to all gardeners since many of our cultivated chrysanthemums and dahlias, for instance, originated in this way. According to the established view, asexually produced offspring are supposed to maintain the identity of the breed. Darwin saw in the existence of bud sports evidence against this view, and for the basic identity of buds and ovules. The conditions of life were in his opinion the cause of both bud and seminal variation though their action in the former case is more direct than in the latter (see Chapter 3). In fact the similarity between the causes of bud and seminal variation is not an external one. It is internal, for both are due to alterations in the constituents of the nuclei. Seminal variations are due to a number of such alterations, of which new combinations of genetic units is peculiar to sexually propagated offspring. This point was recognised by the majority of Darwin's contemporaries, but not by Darwin. His opposition to the distinctive nature of sexual reproduction increased with the years. In his third Notebook on the Transmutation of Species (1838) he spoke of "buds changing into ovules".[8] In the *Origin* (1859) he stated that in the opinion of most physiologists "there is no essential difference between a bud and an ovule in their earliest stages of formation".[9] In 1868 he asserted that sexual and asexual reproduction are essentially the same.[10]

Hybrid Fusion

The fusion of two forms at the cellular and nuclear level can only be achieved by sexual means. The subsequent reversion to the originating forms in the hybrid offspring is also a feature peculiar to seminal hybrids. In the eighteenth and nineteenth centuries opinion was divided as to whether some graft hybrids do or do not display fusion and reversion. Darwin thought that they do and his opinion was repeated by the Lysenko school of biologists quite recently. Darwin cited the relevance of the well-known hybrids known as the Bizzarria Orange and *Laburno-Cytisus adami*. The Bizzarria Orange was raised in 1644 in the Panciatichi Gardens at Florence. In the account given of this plant in the *Philosophical Transactions* in 1675 it was stated to have originated as a scion of orange grafted on to a rootstock of the citron-lemon. From this there emerged a shoot "perfectly retaining the nature and species of both . . .". When it bore fruit it produced on one branch "a mere orange, on another, a citron-lemon, on a third, a citron-lemon-orange, and even sometimes upon one and the same branch all the sorts of this fruit together".[11] Hence it appeared that a union of orange and citron-lemon had been effected by vegetative means—by grafting.

Though this plant has been produced in botanic gardens subsequently it remains a great rarity. *Laburno-Cytisus adami* too is found only in botanic gardens. The first of these "hybrids" was produced by M. Adam of Vitry, near Paris, in 1826. He grafted a portion of the bark of the purple broom *Cytisus purpurea* on to a stock of the common yellow laburnum, *Laburnum vulgare*. One of the shoots issuing from the site of the graft showed characters intermediate between broom and laburnum. When it flowered some of the flowers were intermediate, some appeared exactly like the broom and others like the laburnum. In some flowers, one half was like laburnum and the other half like broom. Hence there seemed to be a tendency for these flowers to revert to the pure parental forms. But intermediacy and reversion were regarded as the distinctive properties of hybrids produced by the sexual fusion of cells, so Alexander Braun, in 1851, suggested that the vegetative cells of the laburnum and the broom had united in a manner analogous to that of the germ cells producing a "true hybrid".[12]

Darwin came to the same unfortunate conclusion, but before we criticise him we should note two facts. First: It was not until 1891 that *Laburno-Cytisus adami* was examined with the aid of a good microscope. Then John Muirhead Macfarlane noted that a "promiscuous mixing of tissue masses" had taken place and further that there was a striking resemblance between the epidermal cells of the graft hybrid and those of the broom. This led Macfarlane to suggest that "the hybrid portion was wrapped round, so to speak, by an epidermis of *C. purpureus*".[13] Orthodoxy, however, prevented him from developing this promising explanation. Instead he went on to assert that a "union of nuclei has taken place",[14] and it was not until 1910 that the two-layer constitution of this graft hybrid was demonstrated by Johannes Buder.[15] It consisted of a core of laburnum tissues covered by an epidermis of broom. No fusion of cells or nuclei had occurred, so it was not really a hybrid at all. In Erwin Baur's terminology of 1909 it was a periclinal chimera.

Of course, Buder could not prove a negative. The

possibility of vegetative fusion still remained. And as late as 1938 a claim was made that such a process had taken place. Hans Winkler claimed he had produced a true intermediate hybrid between the nightshade and the tomato. He supported his contention by citing the chromosome numbers of the hybrid and its parents. That of the "hybrid", at 52 to 56, was approximately intermediate between those of the parents (72 and 24).[16] In 1954 Franz Brabec subjected Winkler's material to a searching analysis and came to the conclusion that this graft hybrid was not a true hybrid but resulted from the production of an octoploid tomato cell ($8x = 96$) the chromosome complement of which had been reduced by subsequent irregular mitoses.[17]

Here we see, that, as in the case of bud variation, the evidence was not correctly interpreted until cytological tools adequate for the task were developed.

Charles Naudin also confused graft-hybrids with seminal hybrids. He believed that hybrids consist of a mosaic of the elements or essences of the two originating species, and these essences struggle incessantly to free themselves from each other and to collect in different parts of the plant. This tendency was supposed to increase with the age of the plant. In *Laburno-Cytisus adami* it takes place in the flowering branches, in the bizarre orange in the fruits and in many hybrids in the petals which, as a result, show stripes of both parental colours.[18] But it is always most active in the pollen and ovules. We know now that except for a few very rare cases it is confined to these two tissues and therefore the results of segregation cannot be seen in the characters of the mother plant, only in those of the offspring. Hence Naudin confused and combined quite different phenomena—vegetative development which gives rise to stripes, variegations and spots, segregation which affects only the progeny, and the growth of "double-tissue" plants or periclinal chimeras. Darwin readily accepted Naudin's confusion of these three phenomena. For him it was yet further evidence for his view of reproduction.

Darwin discussed two other pieces of evidence: the influence of foreign pollen on the tissues of the mother plant, an effect now termed "metaxenia", and the influence of the male agent of fertilisation not only on the form of the resulting progeny but also on the form of those born from subsequent matings, a process now termed "telegony". The occurrence of neither process has been established. Of telegony there was but one reputed case—that of Lord Moreton's chestnut mare, who was mated with a striped-coated Quagga. When the same mare was later mated with a chestnut stallion she bore a foal with a striped coat, thus showing the influence of the father of the previous mating. Karl Pearson writing in 1924 attributed this result either to an impurity in the breed or to "the assertions of kennel men and others endeavouring to screen their responsibility for unplanned matings".[19] In fact the supposed cases of metaxenia and telegony were open to more suspicion than any of the evidence so far discussed.

Today we recognise the twofold distinction between sexual and asexual reproduction which Charles Darwin sought to refute on the basis of these exceptions. The function of bisexual reproduction in producing variation which Erasmus Darwin had stated in 1794 is now universally accepted as its most important role. Gene mutations, it is true, arise independently of sexual reproduction but, unless they are modified, blended and distributed by bisexual reproduction they are invariably of no benefit to the individual or to the species. And in any case, gene mutation is only one source of variation. In sexual reproduction the chromosomes often mutate as a result of crossing over and they are assorted into fresh combinations as a result of meiosis. These changes, too, contribute to the variation upon which natural selection acts. Crossing, far from diminishing variation, increases it.

Now it is clear that Charles Darwin *wanted* to establish the identity of sexual and asexual processes, since so much of his evidence concerned rare phenomena, and it is not difficult to find reasons for this wish. One of the most important reasons is that he inherited a mixed theory of variation from his precursors in evolution. On variation he was "traditional"—he accepted the theory of his predecessors. Thus Buffon, Linnaeus, Lamarck and Blumenbach all attributed variation both to changes in the conditions of life and to cross-breeding. Now changes in the conditions of life affect the organism visibly. New characters are thereby produced which the organism has, so to speak, acquired subsequent to its birth and independent of its heritage. The body cells are affected, but the sex cells are not necessarily affected. Now if sexual and asexual processes are quite distinct from each other it is difficult to see how acquired characters can be inherited by the sex cells and therefore how changes in the conditions of life can give rise to heritable variation in sexually reproducing organisms. By denying any basic distinction Darwin surmounted this problem.

Variation due to Changes in the Conditions of Life

The information available on the effect of the conditions of life was conflicting. So here, as in the case of reproduction, it was difficult to draw a conclusion. It was generally held that animals and plants vary when transported to new conditions of life. The act of domestication or cultivation was looked upon as a form of changed conditions of life since domesticated animals and cultivated plants were known to be more variable than their wild relatives. These facts were favourable to the view that changes in the conditions of life causes variation. Yet, on the other hand, Europeans who migrated to the tropics and lived there for many generations without intermarrying with the indigenous races retained their original skin colour. Negroes did not lose their dark skin colour when they came to live in temperate countries.

Now Blumenbach, the founder of anthropology, believed that the dark-skinned races were derived from the light-skinned races by the action of the strong sunlight of the tropics on man's liver. This action, he believed, had the effect of blackening the bile, and this darkened bile left a carbonaceous deposit in the skin. Thus he was able to explain how it came about that both Europeans and Negroes belong to the same species.

In the early part of the nineteenth century two English surgeon-anthropologists—James Cowel Prichard (1786–1848) and William Lawrence (1783–1867)—attacked Blumenbach for making this statement. They pointed out many facts opposed to it. Lawrence more than Prichard emphasised the fact that it is cross-breeding with other races and not the ancillary change of climate which produce the heritable changes reported.[20] Families which had married into negro families suffered heritable alterations to their skin colour. Families which avoided such inter-marriage retained their original skin colour. But until other writers beside Lawrence and Prichard pointed out marriage as a factor in the situation these reports of changes in skin colour remained conflicting. Thus there were reports of the dark skin colour of European children who had been born in the West Indies and sent to England for their education. It seems very likely that these children were really bastards, the father being a negro. They were retained in the European family and sent with the other children to an expensive English public school where the dark colour of their skin caused comment. Here again inter-marriage had been overlooked, or should we say suppressed, as a factor in the situation!

The belief that domesticated animals and cultivated plants are more variable than their wild relatives was also based on superficial observations. Here again the effects of crossing were overlooked. Lawrence did note that it was animals whose breeding man controlled that varied most and he attributed the many varieties of the pig, sheep, horse, cow, and dog to this cause.[21] But Prichard, after denying that the change of environment causes an inherited darkening of the skin, suggested that this effect of the environment would be seen were it not for the fact that man is protected from local influences by living in houses and adhering to old habits and diets.[22] In coming to this unfortunate conclusion he not only ignored the effects of crossing but also implied that the acquired skin colour is inherited. Previously he had distinguished quite clearly between acquired and congenital characters and had denied the inheritance of the former.[23]

Plants, like animals, were known to vary when transported from their wild habitats to gardens. The question as to how long these variations last, however, remained unanswered until the Austrian botanist, Anton Kerner von Marilaun, carried out his famous transplant-experiments during the years 1875–1880. He grew lowland plants at two alpine stations, 2,195 m. and 1,215 m., and at two lowland stations, the Innsbruck and Vienna botanic gardens, 569 m. and 180 m. There was a marked reduction in height, number of internodes and size of plant organs with increasing altitude. But progeny raised in the botanic gardens from seeds harvested at the alpine stations showed no trace of the acquired alpine characteristics. There was no escaping the conclusion that "in no instance was any permanent or hereditary modification in form or colour observed".[24]

Experiments of a similar nature had been carried out by Andrew Knight, President of the Horticultural Society of London. As early as 1797 he reported the results of experiments on the effect of rich garden soil on fruit trees. He planted fruit trees in clay and garden soils and compared the growth of scions taken from them and grafted on to trees growing in poor soil. Despite the greater luxuriance of the trees growing in garden soil, he observed no difference in the branches and leaves of the grafted scions taken from them. The luxuriance of growth which characterised their parents ceased when the conditions producing it were taken away. So it seemed to Knight that the many attempts of his forefathers to improve cultivated plants by growing them in rich soils were all useless. Their labours, he said,

> here began where they might as well have ended, . . . no permanent change can be made in the future produce of the seeds by any mode of cultivation which can be adopted subsequent to their being taken from the parent tree.[25]

But Knight's experiments went unheeded. As a result the important role of cross-breeding was once again submerged beneath a confused mass of evidence for the effects of the conditions of life which lead to variability.

Mendel was of the same opinion as Knight regarding the effects of cultivation. He was also able to give a precise explanation of the variability of cultivated plants on the basis of segregation and the independent assortment of characters. This important passage from his paper on *Pisum* is reprinted in the Appendix. The factor, he said, which "so far has received little attention", is crossing. This process is facilitated by the fact that "cultivated plants are mostly grown in great numbers and close together . . .[26] Darwin was almost brought to the same conclusion in 1881 when he read Hermann Hoffmann's review of his experiments to test the effects of growing plants in unnatural conditions. So negative were the results that Darwin was "staggered". When he read further he was amazed to find that:

> Hoffmann even doubts whether plants vary more under cultivation than in their native home and under their natural conditions. If so, the astonishing variation of almost all cultivated plants must be due to selection

and breeding from the varying individuals. This idea crossed my mind many years ago, but I was afraid to publish it, as I thought that people would say, "how he does exaggerate the importance of selection".[27]

Pangenesis

In 1865 Darwin worked out a hypothesis to explain how the effects of the conditions of life can be transmitted to the germ cells, and he called it "Pangenesis". It was published in vol. 2 of his book *The Variation of Animals and Plants under Domestication*. There is also a manuscript in the Cambridge University Library bearing the title "Hypothesis of Pangenesis". This would appear to be the draft which he made in 1865 and sent to Huxley for comment. It shows much more clearly than does the published account how Darwin arrived at the hypothesis in an inductive manner starting from the assumption of the identity of all forms of reproduction. The most important part of this manuscript is reproduced in the Appendix. Here we will summarise Darwin's argument under four heads.

(1) Since the buds of a tree and the polypi of a coral have an independent as well as a communal life one may regard the organism as being composed of numerous semi-independent units. This led Darwin to a particulate theory.

(2) Since a portion of a planarian and the buds of a tree can regenerate the whole organism, the protoplasm for making the whole organism must be present in all parts of the body and not merely in the germ cells.

(3) When a polyp buds off another polyp it is producing a fresh individual from the protoplasm surplus to its own requirements. Therefore all the parts of an organism have surplus protoplasm which they can "throw off".

(4) Double monsters are always united by like members and not by dissimilar members thus showing that like parts have an "affinity" for each other. The units

thrown off from the various parts of the body also have a mutual affinity which leads to their accumulation in buds and in the sexual elements.

Now pangenesis is really a version of the ancient Hippocratic doctrine of the formation of the seed by the body tissues which was re-stated by numerous authors after Hippocrates. Darwin could have borrowed the idea from any of these and maybe he did so, even if unconsciously. Nevertheless I still think that the hypothesis represents the crystallisation of Darwin's thoughts over a period of a quarter of a century, thoughts which began with his wonder at the ability of a planarian to regenerate after division.[28] This conclusion is supported by Darwin's remark to Charles Lyell in 1867 that Pangenesis "is 26 or 27 years old",[29] thus taking us back to 1841–2 when he was reading the famous 2-volume *Physiology* of Johannes Mueller. And sure enough, Darwin's copy of this work contains marginalia which bear directly on pangenesis and concern the affinity between like parts of double monsters and budding as a process in which superfluous material is separated from the organism. In his index to these passages at the back of the book he wrote, though probably at a later date, the word pangenesis.[30]

Pangenesis was the result of Darwin's favourite pursuits: seeking analogies between apparently unrelated phenomena and framing hypotheses to account for them. Thus he told Bentham that the launching of Pangenesis had been a great relief to him because "I could not endure to keep so many large classes of facts all floating loose in my mind without some thread of connection to tie them together in a tangible method".[31] We have seen that the most important step which he took in this direction was his denial of the distinction between sexual and asexual reproduction. Most of the evidence for this view concerned rare phenomena, but in Darwin's day it could not be reconciled as it can today with a belief in the distinctive nature and role of sexual reproduction.

NOTES

(1) Darwin, E. 1794. *Zoonomia; or, the laws of organic life.* Vol. i, p. 487.

(2) Dzierzon. 1845. (Title of article unknown.) Bienenzeitung, herausgegeben von Dr. C. Barth und A. Schmidt in Eichstadt. Jahr 1, p. 113.

(3) Dzierzon. 1854. "Die Drohnen." Der Bienenfreund aus Schlesien Brieg. No. 8, pp. 63–64. This passage was translated by Conway Zirkle and is reprinted in the appendix.

(4) Siebold, C. T. E. von. 1857. *On a true parthenogenesis in moths and bees; a contribution to the history of reproduction in animals.* Trans. W. S. Dallas. London. p. 12.

(5) Ibid., p. 106.

(6) Smith, J. 1839. "Notice of a plant which produces perfect seeds without any apparent action of pollen. J. Linn. Soc., *18*, 511.

(7) Millardet, M. A. 1894. "Note sur l'hybridation sans croisement ou fausse hybridation." Mém. Soc. Sci. phys. nat. Bordeaux, sér. 4, *4*, 365.

(8) N3, p. 137.

(9) O1e, p. 8.

(10) VuD, vol. ii, p. 365.

(11) Natus, P. 1675. "A phytological observation concerning oranges and lemons, both separately and in one piece produced on one and the same tree at Florence." Phil. Trans., *9*, No. 114, 314.

(12) Braun, A. 1853. "The phenomenon of rejuvenescence in nature, especially in the life and development of plants." (Trans. A. Henfrey from the German of 1851) *Botanical and Physiological Memoirs published by the* Ray *Society.* London. p. xxii, 316–317.

(13) Macfarlane, J. M. 1895. "A comparison of the minute structure of plant hybrids with that of their parents, and its bearing on biological problems." Trans. Roy. Soc. Edinb., *37*, 268.

(14) Ibid., p. 269.

(15) Buder, J. 1910. "Studien an *Laburnum adami.* I. Die Verteilung der Farbstoffe in den Blütenblättern." Ber. dtsch. bot. Ges., *28*, 188–192.

(16) Winkler, H. 1938. "Ueber einen Burdonen von *Solanum lycopersicum* und *Solanum nigrum.*" Planta, *27*, 684.

(17) Brabec, F. 1954. "Untersuchungen über die Natur der Winklerschen Burdonen auf Grund neuen experimentellen Materials." Planta, *44*, 601.

(18) Naudin, C. 1863. "Nouvelles recherches sur l'hybridité dans les végétaux." Ann. Sci. Nat. Botanique, sér. 4, *19*, 192–193.

(19) LFG, vol. ii, p. 159.

(20) Lawrence, W. 1819. *Lectures on physiology, zoology and the natural history of man.* London. pp. 97 and 540.

(21) Ibid., p. 263.

(22) Prichard, J. C. 1826. *Researches into the physical history of mankind.* 2nd ed., London. Vol. ii, pp. 582–583.

(23) Ibid., vol. ii, pp. 544–545.

(24) Kerner von Marilaun, A. 1895. *The natural history of plants, their forms, growth, reproduction, & distribution.* Translated and edited by F. W. Oliver. 2 vols. London. Vol. ii, p. 514.

(25) Knight, T. A. 1797. *A treatise on the culture of the apple and pear and on the manufacture of cider and perry*. Ludlow. pp. 32–33.

(26) MRI, p. 32.

(27) Darwin, F. (Ed.). *The life and letters of Charles Darwin, including an autobiographical chapter. Edited by his son, Francis Darwin.* 3 vols. London. Vol. iii, p. 345. (Referred to in future as LLD.)

(28) Darwin, C. 1845. *Journal of researches into the natural history and geology of the countries visited during the voyage of H.M.S. "Beagle" round the world, under the command of Capt. Fitz Roy R. A.* London. 2nd ed., p. 27.

(29) LLD, vol. iii, p. 72.

(30) Mueller, J. 1838–1842. *Elements of physiology. Trans. W. Baly.* 2 vols. London. Vol. ii, annotations on the rear flyleaf.

(31) MLD, vol. ii, p. 371.

SUGGESTIONS FOR FURTHER READING

Brabec, F. 1965. "Pfropfung und Chimären, unter besonderer Berücksichtigung der entwicklungsphysiologischen Problematik." *Handbuch der Pflanzenphysiologie.* Herausgegeben von W. Ruhland, vol. xv, pt. 2, 388–479.

Darlington, C. D. 1958. *Evolution of genetic systems.* 2nd ed. Oliver & Boyd, Edinburgh and London; 1958, Basic Books, New York.

Ephrussi, B. et al. 1964. "Hybridization of somatic cells *in vitro*." Symposia of the International Society for cell Biology, vol. iii, *Cytogenetics of cells in culture.* Academic Press, New York and London, pp. 13–25.

Gustafsson, A. 1945. "Apomixis in higher plants. Part 1. The mechanism of apomixis." Acta Univ. lund., *42*, No. 3, 1–370.

Haskins, C. P. and E. F., and Hewitt, R. E. 1960. "Pseudogamy as an evolutionary factor in the poeciliid fish *Mollienisia formosa*." Evolution, *14*, 473–483.

Poulton, E. B. 1908. *Essays on evolution 1889–1907.* Clarendon Press, Oxford.

Tilney-Basset, R. A. E. 1963. "The structure of Periclinal Chimeras." Heredity, London, *18*, pt. 3, 265–285.

Chapter Five

Gregor Mendel

To this day an air of mystery surrounds the name of Gregor Mendel,the quiet Moravian monk who laid the foundations of genetics. How did he do it and why was his work ignored? It was to answer questions such as these that William Bateson (1861–1926) went to Brno in the winter of 1904–1905, but he found nothing. Apparently Mendel's successor, Abbot P. Anselm Rambousek, burnt all his private papers. Fortunately Mendel's official documents remained in the monastery archives, where they were later discovered by a young priest by the name of P. Anselm Matousek. At the time of Bateson's visit Matousek was studying to enter the priesthood, but as soon as he had been ordained in 1906 he set to work to collect relics and manuscripts relating to Mendel. In this work he was joined by Hugo Iltis, a young professor from the Brno Gymnasium. The material which these men unearthed has miraculously survived two world wars and the subsequent political and scientific upheaval in Czechoslovakia. Now it has been carefully arranged together with fresh material in the Mendel Memorial Hall by the staff of the Gregor Mendel Department of Genetics of the Moravian Museum. There one can see the answers which Mendel wrote to his examination questions, his brief autobiography, his school and university reports and his letters to Naegeli, to name just the more important items. But despite all that has been discovered and preserved we have no direct information on the sources of Mendel's inspiration, nor have we as much information as we would like on the reasons for his failure to attract the interest of any other scientists of his time.

We have only one first-hand account of Mendel which includes a reference to his researches on *Pisum*, and that was supplied by a horticulturist named C. W. Eichling in an article he wrote in 1942, when he was 86 years of age. In the summer of 1878 he was on a business trip in Central Europe visiting nurserymen and hybridists as the representative of the firm of Louis Roempler, of Nancy. In Erfurt he visited Ernest Benary who was an experienced hybridist and knew of Mendel's breeding experiments. He advised Eichling to visit Mendel if he was passing through Brno. This he did and he has now left us a touching account of the warm reception which the 56-year-old abbot gave to him, a mere 22-year-old business representative. Mendel gave him lunch and showed him the monastery gardens including

> ... several beds of green peas in full bearing, which he said he had reshaped in height as well as in type of fruit to serve his establishment to better advantage. I asked him how he did it and he replied: "It is just a little trick, but there is a long story connected with it which it would take too long to tell." Mendel had imported over 25 varieties of peas, which shelled out readily, but did not yield very well because some of them were bush types. As I recall it, he said that he crossed these with his tall local sugar-pod types, which were used at the monastery. I told Mendel that I had promised to make a report to Benary regarding these experiments, but Mendel changed the subject and asked me to inspect his hothouses.[1]

It was unfortunate that Eichling was out of touch with scientific problems and did not know how hard-pressed were biologists at that time to find the laws of inheritance and explain the behaviour of hybrids. As it was, Mendel was easily able to change the subject and Eichling, no doubt out of courtesy, did not raise the matter again. Clearly Mendel was very sensitive about his work on *Pisum*. This is to be expected since his attempt to interest Carl Naegeli had met with little short of a snub. Eichling tells us that when he inquired about Mendel in the town his informant replied that "while der Herr Abt was one of the best-beloved clerics in Brünn, not a soul believed his experiments were anything more than a pastime, and his theories anything more than the maunderings of a charming putterer".[2] This seems to have been the attitude of informed scientific circles also. To plant orchards, graft fruit trees and keep bees were normal occupations, but counting tens of thousands of round and wrinkled peas was distinctly abnormal, a point of view which some biologists still take today. The reasons for this attitude will be discussed in the last chapter. Here we are concerned more with the sources of Mendel's inspiration. Of course we cannot expect to find a straight answer to this question, but by reviewing the details of Mendel's background and education some clues ought to emerge.

Mendel's Life

Mendel's parents were peasant farmers in the Silesian village of Heinzendorf, now bearing the Czech name Hynčice. There were five children, four daughters and one son. Two daughters died leaving Veronica, born 1820, Johann, born 1822 and renamed Gregor in 1843, and Theresia, born 1829.

At the village school in Heinzendorf Mendel's exceptional ability soon became apparent and in 1833 on the advice of the schoolmaster Thomas Makitta and the pleadings of Mendel his parents agreed to transfer him to the Piarist High School in Leipnik (Czech Lipnik). Within a year he had moved on again, this time to the Gymnasium in Troppau (Czech Opava). There he stayed until the end of his school days in August 1840, when he received his leaving certificate which attests to his great industry and all-round ability for it is full of "*I ems*", i.e. *prima classis cum eminentia.*

For the continuation of his studies he went to Olmütz (Czech Olomouc) and enrolled in the philosophy course which was held at the University Philosophical Institute. In his autobiography Mendel describes his time in Olmütz as follows:

> . . . his first care was to secure for himself the necessary means for the continuation of his studies. Because of this, he made repeated attempts in Olmütz, to offer his services as a private teacher, but all his efforts remained unsuccessful because of lack of friends and recommendations. The sorrow over these disappointed hopes and the anxious, sad outlook which the future offered him, affected him so powerfully at that time that he fell sick and was compelled to spend a year with his parents to recover.[3]

When he had recovered, his sister Theresia gave him a part of her dowry so that he could return to Olmütz. This he did in 1841 when he also found tutorial work and was able to stay long enough to complete the first two years in philosophy. But such was Mendel's industry and devotion to his studies and conscientiousness as a tutor that his health broke down again and again and he "felt compelled to step into a station of life, which would free him from the bitter struggle for existence".[4] Even at Troppau he had had difficulties since his parents had paid a reduced fee which meant that he was put on "half rations", his bread and butter being supplied by his parents in Heinzendorf twenty miles away. In 1843 his difficulties became acute. Five years before, his father had suffered chest injuries when the trunk of a tree fell on him, and because he was no longer fit to manage the farm he sold it to his son-in-law, Alois Sturm, in 1842. In the contract Mendel's father made provision for his son, including payment of the expenses connected with the first mass, should Mendel enter the priesthood. Hence it seems that the family had this career already in mind for their son before he finished his studies in Olmütz. In 1843 Mendel discussed his future with Friedrich Franz, Professor of Physics at Olmütz, and he recommended that Mendel should go to the Augustinian Monastery in Brno, the city in which Franz had taught for nearly twenty years before coming to Olmütz. Professor Franz had, in fact, lived in the monastery and had just been asked to select candidates to go there in 1843. He chose Mendel whom he described as "a young man of very solid character. In my own branch he is almost the best".[5]

Mendel and his family acquiesced in the Professor's suggestion and on October 9th, 1843, he was admitted to the monastery as a novice under the name of "Gregor". From his autobiography it is clear that he did not, like many clerics, feel called to the Church, but as he said himself, "his circumstances decided his vocational choice"; and from the time of his entry to the monastery he served it loyally. In return he was freed from financial troubles and he found himself at the centre of Moravian culture and scientific study. Many of the members of the monastery were full-time teachers either at the Philosophical Institute in Brno or at the Gymnasium. Some of them stayed in the monastery for a few years and then left to take up professorial chairs in universities. Friedrich Franz, for instance, went to Olmütz as professor of physics. Hence it must not be assumed that in entering the Augustinian Order in Brno Mendel was cutting himself off from the world and from cultural and scientific developments. On the contrary his situation was quite the reverse.

In his school days Mendel's chief difficulty had been his intensely nervous disposition. Under conditions of stress in his studies his health had broken down at least four times and as a priest this feature of his constitution was also manifest when he visited the beds of the sick. So violent was his reaction that Prelate Napp decided to relieve him of all pastoral duties. Apparently he was siezed with an insurmountable fear at the sick-bed and the sight of invalids, and he became dangerously ill. Instead he was appointed supply teacher to the Gymnasium in Znaim (Czech Znojmo). He started on October 7th, 1849, and he soon proved a great success. There is no doubt that Mendel thoroughly enjoyed his life as a teacher and that he was popular both with staff and pupils. So it was natural that he should wish to receive a permanent appointment, but for this he had to take the examination for Gymnasium teachers. Arrangements were therefore made for Mendel together with two other supply teachers at the Gymnasium to be examined in the summer of 1850.

The examination was in three parts, first the candidates had to write two essays during the summer term. If these were satisfactory they were allowed to proceed to the second and third parts of the exam held in Vienna and consisting of a *viva* followed by an unseen paper. For the Professor of physics in Vienna, Baron von Baumgartner, he had to write an essay on the chemical and physical properties of air, and for the Professor of zoology, Rudolf Kner, an essay on igneous and sedimentary rocks. Mendel's essays are preserved at Brno to this day. They show the clear exposition and logical precision which is characteristic of Mendel's later work. Baumgartner was well satisfied with Mendel's essay on air, but Kner was dissatisfied with his essay on geology. One or two of the comments which Kner made at the side of Mendel's text show that he had little grounds for criticism. Why, for instance, did he quibble with the statement that: the cosmological theories of Kant and Laplace provide a "simple and satisfactory explanation of almost all geological phenomena", or that: "The process of carbonisation is probably the outcome of a defective process of fermentation, beginning in driftwood that has accumulated in a lowland swamp, and proceeding up to a certain point, but then ceasing because the access of air is inadequate". However, Kner allowed Mendel to continue the exam.

Mendel was instructed to come to Vienna for the second part of the exam on August 1st. A further letter telling him to wait until the next academic year did not reach him so that when he presented himself at Professor Baumgartner's office he was not expected. Evidently he succeeded in persuading the Professor to hold the exam for Mendel was that day given the written paper followed a fortnight later by the *viva-voce* exam. Again Baumgartner was satisfied with him and Kner was not, but this time Kner had good reasons. Kner had set him a question on the classification and uses of the *Mammalia* to which Mendel gave a very poor answer. He just wrote down whatever came into his head. At the *viva* he gave a more satisfactory impression. In October he was informed that he had failed but was advised to sit the exam again "after the lapse of not less than one year". Mendel must have been very disappointed that autumn, but not entirely without hope for he had made a very good impression on Baumgartner, and when, a year later, Prelate Napp wrote to the Professor to ask why Mendel had failed, Baumgartner replied advising the Prelate to send Mendel to the University of Vienna to obtain a more thorough grounding in natural science. Meanwhile Mendel had distinguished himself once more as an educator, this time as supply teacher at the technical high school in Brno.

Student in Vienna

We come now to what was undoubtedly the most formative period in Mendel's life, his time in Vienna as a student in the philosophical faculty of the University from October 1851 to August 1853. Mendel had thriven in the intellectual climate of Brno and now at Vienna one can imagine with what intense interest he must have listened to such men as Franz Unger, Andreas von Ettinghausen and Christian Doppler, for these men were no mere dispensers of orthodox knowledge, but students of science alive to current problems in their subject; men whose enthusiasm was infectious.

Franz Unger (1800–1870) was Professor of Plant Physiology at the University in Vienna from 1849–1866 and he taught in both high school and University. Apparently he was not a brilliant speaker, but he knew how to captivate his students by expressing his own enthusiasm for science. His ability as a teacher and his approachable and amiable nature won for him such devotion from his students that they rose as one man in his defence when he was attacked from clerical quarters, and called a corrupter of youth who ought to be dismissed from the University. This charge arose out of the publication of his popular *Botanical Letters* (German edition 1852, English edition 1853) in which he denied the fixity of species and asserted that the plant world "has gradually developed itself step by step".[7]

We have no direct information as to the sort of instruction in cytology which Mendel received at Vienna but we can get some idea of what Unger would have emphasised in his instruction from his *Botanical Letters*, his textbook *Anatomie und Physiologie der Pflanzen*, and his letters to Stephen Endlicher which were published by Haberlandt in 1899. Thus in his *Botanical Letters* he made the cell the theme of all his discourses. He called it the "factotum", the "Proteus" which determines all the higher unities. The mystery of the origin of cells, he said, "consists in the fact that the plant develops each one that she employs from others previously existing".[8] As to the scientific basis of the architecture of plants he said:

> The man who has hitherto penetrated into this obscure region of research, who has attempted to examine stone by stone in their production and application, who has

given us both ground plans and elevations of some plant structures in which each element is marked with the number its architect intended for it, this man is Carl Naegeli.[9]

In his letters to Endlicher he spoke with admiration of both Schleiden and Naegeli. Of the former he said: "We have long needed a man like him. In our science it is not we, but he, who is opening a new epoch."[10] He was referring, of course, to the study of cells. Schleiden in his famous textbook, the *Grundzüge*, had expressed his aim as:

> . . . to establish the necessity of embracing, as a fundamental principle in the study of the whole, the existence of an essential life in each separate cell. Hence arises the necessity for carrying on investigations in the first instance in the individual cells, or in portions of the vegetable structure, in which we have to do with few cells in combination. On these we must make our first experiments, and from them draw our first conclusions, which may then proceed to apply to subsequent investigations into the general structure of plants.[11]

When Naegeli went to Jena in 1842 he worked with Schleiden on the cellular basis of growth. His discovery of perpetually dividing cells at the apex of shoots and roots and of fixed patterns of cellular differentiation caused considerable interest. Unger referred to these discoveries in the above quotation from his *Botanical Letters* as "ground plans and elevations". Hence Mendel must surely have learnt of Naegeli's important work when he was at Vienna.

Unger also included an account of all the latest researches into the fertilisation process up to 1855, in his textbook of that year, and most important of all, he followed this account with a brief summary of the results of hybridisation experiments most of which he gleaned from the works of Gaertner. The books written by Gaertner and Koelreuter are all cited in Unger's bibliographies in this textbook. So it is highly probable that it was from Unger and at Vienna that Mendel also learnt of the work of his precursors.

We have already mentioned the fact that Unger was nearly suspended for expressing evolutionary views in his *Botanical Letters*. The German edition of these letters appeared in 1852. Mendel was taught by Unger from October 1852 to April 1853, so he must have been familiar with his views. Mendel also visited Vienna in 1856, so he may have heard, too, about the attempt to make Unger resign that year.

In his textbook of 1855 Unger rejected, once more, the belief in the stability of species. Instead he held that variants arise in natural populations and that the slight variants give rise to varieties and sub-species whilst the larger variants form specific differences. This view was at variance with the opinion of Koelreuter and Gaertner and it would appear that Mendel carried out his experiments in order to decide the issue. He realised that it was an arduous labour but he believed it was "the only right way by which we can finally reach the solution of a question the importance of which cannot be overestimated in connection with the history of the evolution of organic forms".[12]

The task which Mendel had set himself was arduous because it necessitated a determination of "the number of different forms under which the offspring of hybrids appear", or a definite ascertaining of "their statistical relations".[13] This is unquestionably the most novel part of Mendel's approach to hybridism. Where did he obtain his knowledge of statistics? It is far more likely that he obtained it from his physical studies than from his biological studies, and it should be remembered that his ability as judged by the yardstick of examinations was greater in physics than it was in biology. He was at home in the more mathematical sciences. Christian Doppler, the discoverer of the "Doppler effect", taught him experimental physics. He had been Director of the Physical Institute and Professor of experimental physics in Vienna since 1850. When Mendel came in 1851 Doppler was 48 years of age and at the height of his powers. His chief interest was in the mechanics of electromagnetic waves and in light waves, but he was also interested in more general mathematical and geometrical problems.

Andreas von Ettinghausen, who taught Mendel higher mathematical physics and the use of physical apparatus was, like Doppler, interested in the mathematical analysis of physical problems, chiefly those concerning wave mechanics. If one glances through the titles of the papers published by Ettinghausen and Doppler which are listed in the Royal Society *Catalogue of Scientific Papers* (1800–1863), vol. ii (1868), one cannot fail to be impressed by one feature common to almost all of them—emphasis on the mathematical approach. We can say then, that if Mendel gained his statistical knowledge in Vienna he must surely have gained it from Ettinghausen or Doppler.

For a time Mendel acted as demonstrator at the Physical Institute in Vienna. This suggests that he would have had ample opportunity for becoming familiar with the physicist's approach to experiment which, unlike that of the naturalist, is not to make many observations in a Baconian manner and then to seek for an underlying pattern but to analyse a problem first, arrive at a solution on paper and only then to carry out an experiment to confirm or refute the solution. With this aim in mind suitable experimental material is sought and fitting experiments are planned. The problem is, of course, suggested in the first place by observational data; e.g. the interference and diffraction of light are observed when oil is poured on water and when the sun shines through a hazy sky. The physicist has then to arrive at a concept of light which can account for these phenomena. He postulates, let us say, longitudinal waves, and designs experiments to verify or deny his wave theory.

Franz Unger no doubt described the facts known at the time about hybrids—their uniformity in the F1 generation and their tendency to reversion in the F2 generation. Mendel in his paper of 1866 took up the subject from this point. The facts did not need re-establishing, they called for explanation,* and in 1866 no one had produced a satisfactory solution. The similarity between these two problems—concerning light and hybrids—lies in the approach needed to solve them. It seems that Mendel became familiar with the correct approach from his studies in physics at Vienna and not from his studies in biology.

No information is available on Mendel's activities immediately after his return to Brno at the end of July 1853, but in May 1854 he was once more appointed supply teacher, this time at the newly-founded Oberrealschule in Brno. For the next sixteen years he taught physics and natural history and was form master of the second class. When Iltis was collecting material for his biography of Mendel in the early part of this century he found many former pupils of Mendel still living in Brno. They all spoke warmly of the cheerful "rather stocky" cleric who was kind, conscientious, just and above all a good teacher.

In 1855 Mendel once more applied to take the teachers' examination at Vienna. In May 1856 the exam took place, but this time he could not even complete the written paper. After answering the first question he became so ill that it was impossible for him to write any more. He retired from the exam and returned to Brno. Soon after this incident his father and his uncle made the long journey from Silesia to Brno to see him. It was the only time they ever came. Evidently Mendel was very ill and it was presumably the same illness which had plagued him at times of stress when he was a schoolboy and a

* Mendel had already observed F1 uniformity in the course of his work as a plant breeder.

student. Dr. Joseph Sajner, who has made a special study of Mendel's illness, calls it an "unstable psychological constitution".[14] Mendel had already been ill earlier in the year, no doubt owing to the stress of his studies, and when faced with the "Vienna ordeal" again his nerves gave way. He never sat the exam again, but remained a supply teacher until the close of his teaching career in 1868 when he became Abbot.

Research Period

Fortunately Mendel recovered sufficiently to make a start on his experiments with *Pisum* in the summer of 1856. He had already tested 34 varieties of the edible pea for purity of type and suitability as research material, for, as he remarked: "The value and utility of any experiment are determined by the fitness of the material to the purpose for which it is used . . ." This statement comes after his explanation of the aims of the work—to find out the statistical relations of the various hybrid offspring. It is the physicist's approach, and as the plan of the experiments unfolds one becomes more and more convinced that as Sir Ronald Fisher suggested in 1958 "the experiments were in reality a confirmation, or demonstration, of a theory at which he had already arrived . . .".[15] From the testing of the 34 varieties he selected 22 for the experiments which followed—the seven series of unifactorial crosses by which he established the 3 : 1 ratio, followed by the two bifactorial and the one trifactorial crosses by which he established the independent segregation of character elements. Then he rounded off the series by devising bi- and tri-factorial crosses to test predictions based on his theory of factorial inheritance.

The whole theory rests on one inference which no one else had thought of making. It was simply the prediction of the number of different forms that would result from the random fertilisation of two kinds of "egg cells" by two kinds of pollen grains. Naudin had postulated the segregation of specific essences in the formation of germ cells; Mendel postulated the segregation of character elements. Thus if the character difference between the two forms crossed be represented by the letters *A* and *a*, then following Mendel's explanation, the resulting hybrid *Aa* forms not one type of germ cell *Aa* only, but two types—*a* and *A*, and the following fertilisations are possible:

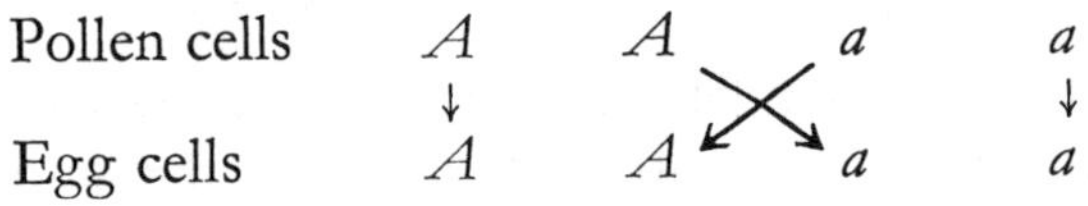

Pollen cells	*A*	*A*	*a*	*a*
	↓	↘	↙	↓
Egg cells	*A*	*A*	*a*	*a*

$$\text{giving} \qquad \frac{A}{A} + \frac{a}{A} + \frac{A}{a} + \frac{a}{a}$$

or: 1 pure *A*; 2 hybrid *Aa* : 1 pure *a*, and where *A* is dominant over *a* the second class appears like the first giving 3*A* : 1*a*. All the other ratios are based on these two, with the exception of back-cross ratios.

By way of demonstration Mendel obtained results which Sir Ronald Fisher thought too good to be true.[16] Thus for the character differences round seeds to wrinkled and yellow to green Mendel counted 5,474 round : 1,850 wrinkled and 6,022 yellow : 2,001 green, which are in the ratio 2·96 : 1 and 3·01 : 1 respectively. The subject of the reliability of these results is discussed in the appendix to this chapter. Here we will only note that Fisher's conclusion applies with equal force to Tschermak's results, and that there is another and more plausible explanation of this fact than the one given by Fisher.

Mendel concluded his experiments with *Pisum* in 1863, but knowing that the results he had obtained were not easily compatible with "contemporary scientific knowledge, and that under the circumstances publication of one such isolated experiment was doubly dangerous; dangerous for the experimenter and for the cause he represented", he made "every effort to verify, with other plants, the results obtained with *Pisum*".[17] First he crossed the French bean (*Phaseolus vulgaris*) with the bush bean (*P. nanus*) and obtained clear 3 : 1 ratios between the segregates for three pairs of contrasted characters. Then he crossed the bush bean with the scarlet runner (*P. multiflorus*) and again obtained 3 : 1 ratios. Unfortunately he did not give the numbers on which these ratios were based. The two parental flower colours, white (*P. nanus*) and red (*P. multiflorus*), however, did not segregate according to the 3 : 1 ratio. He suggested that two independently acting colours may be involved here, in other words that it is a case of interaction between two pairs of contrasted characters; nevertheless he had not a sufficiently large number of hybrid offspring to prove his point. A manuscript in Mendel's handwriting which is thought to relate to colour inheritance in *Phaseolus* is described in the appendix. This character difference seems to have been the first which Mendel was unable to fit into his theory. It was in view of this failure that he discussed his work on *Pisum* at two meetings of the Naturforschenden Vereins in Brno (February 8th and March 8th) "in order to inspire some control experiments".[18] He encountered "as was to be expected, divided opinion; however", he said, "as far as I know, no one undertook to repeat the experiments",[19] but the Society did ask him to publish his lecture in 1866.

On the first page of the manuscript of Mendel's paper on *Pisum* is written "40 Separatabdruck", i.e. 40 reprints. We do not know whether he sent out 40 but only four of them have so far been discovered. He sent one to Naegeli, one to Anton Kerner von Marilaun (1831–1898) at Innsbruck, one to an unknown recipient which found its way into the library of Martius Wilhelm Beijerinck who sent it to Hugo de Vries, and one to an unknown destination which turned up in the library of Theodor Boveri (1862–1915), was given to the Kaiser Wilhelm Institut für

Biologie in Berlin and is now in the Max Planck Institut in Tübingen. One hundred and fifteen copies of the journal were sent out, 12 to local addresses, 8 to Berlin, 6 to Vienna, 4 to U.S.A., and 2 to Great Britain (the Royal Society and the Linnaean Society). It is notorious that Mendel failed to arouse any interest in the local Society or in any of the institutions to which the journal was sent. Only in Naegeli did he find a correspondent ready to discuss his work.

Correspondence with Naegeli

The correspondence between Mendel and Naegeli lasted from 1866 to 1873. Naegeli had published a critical review of the work of Koelreuter and Gaertner, so he could hardly fail to be interested in Mendel's experiments. Unfortunately he could not believe that Mendel's explanation was the right one because he was quite sure that all the offspring of hybrids are variable. There cannot exist true-breeding offspring. Naegeli's opinion of Mendel's so-called constant forms in the F2 generation was that "they would sooner or later be found to vary once more. A, for instance, has *a* in its body, and when inbred cannot lose that element".[20] In Chapter 4 we have seen that in the genus *Hieracium* this perpetual hybridity is common owing to apomixis. Since Naegeli was working with *Hieracium* he had good grounds for his belief but he did not know that he was dealing with an exceptional case whereas Mendel, working with *Pisum*, was studying the normal case. Unfortunately Naegeli encouraged Mendel to concentrate his attention on *Hieracium*, and of course he failed to find any agreement with *Pisum* although he hybridised them for five years (1866–1871). Indeed he experienced great difficulties in producing any hybrids at all, and the majority of those he raised bred true much to his surprise. In 1869 he read a paper on *Hieracium* to the Brno Society for the Study of Natural Science. He referred to Max Ernst Wichura (1817–1866), who had reported the production of true-breeding willow hybrids, and said:

> In *Hieracium* it might appear that we have to do with an analogous case. As to whether the polymorphism of *Salix* and *Hieracium* may be connected in some way with the peculiar behaviour of their hybrids, this is a question which as yet we can only raise but cannot answer.[21]

What Mendel called "peculiar behaviour" was apomixis, and today we know that hybridisation followed by apomixis does lead to the production of different true-breeding forms. Here again we see Mendel realising the evolutionary significance of his results, although in this case he did not know the cause.

In addition to *Hieracium* and *Salix* there had been other reports of other true-breeding hybrids. Gaertner listed nine;[22] one of these was *Geum urbanum* × *rivale*. It was thought to be identical with the naturally occurring avens *Geum intermedium*, so in 1866 Mendel repeated Gaertner's cross and he confirmed that the resulting F1 hybrids were identical with *G. intermedium*. In the summer of 1867 he raised F2 and back-crossed plants which presumably came into flower in 1868, but we have no information about them because Mendel stopped reporting on them after November 1867.[23] In fact neither the naturally occurring nor the artificially produced *Geum intermedium* breeds true. Segregation occurs, but only in the character of anthocyanin pigment has a clear 3 : 1 ratio been demonstrated.[24] It seems that the majority of character differences between the wood and water avens are determined by more than one gene and that polyploidy complicates segregation. Some notes which Mendel made on these hybrids are discussed in the Appendix.

Abbot of the Monastery

In the spring of 1868 Mendel was appointed Napp's

successor as Abbot of the Monastery. In the following May he wrote to Naegeli and told him about his promotion:

> Recently there has been a completely unexpected turn in my affairs. On March 30 my unimportant self was elected life-long head, by the chapter of the monastery to which I belong. From the very modest position of teacher of experimental physics I thus find myself moved into a sphere in which much appears strange to me, and it will take some time and effort before I feel at home in it. This will not prevent me from continuing the hybridisation experiments of which I have become so fond; I even hope to be able to devote more time and attention to them, once I have become familiar with my new position.[25]

Unfortunately Mendel never found the leisure time which he had hoped for. Consequently his hybrid studies came to an end in 1871. His remaining years were made difficult partly by his obstinacy in refusing to pay the new ecclesiastical tax and partly by ill health. The aim of this tax was chiefly to raise the funds for augmenting the stipends of parish priests. It was a heavy tax—10 per cent of the value of the entire property of the monastery—but there were a number of "softening clauses" by which the valuation could be reduced. Mendel ignored these, sent in his valuation of the monastery as 777,511 florins and then refused to recognise the legality of the tax. The ensuing controversy, which was not settled until after Mendel's death, harmed relations between the monastery and the civic authorities and saddened Mendel. At the same time Mendel's health declined. He became a heavy smoker, often smoking as many as twenty cigars a day, and his nephew Alois Schindler records the fact that Mendel's pulse rate was frequently as high as 120.

Alois and Ferdinand Schindler were the sons of Mendel's favourite sister Theresia. She had sacrificed a part of her dowry in the cause of Mendel's education. Mendel more than repaid his sister for her kindness by giving her sons a home in the monastery while they were attending the Gymnasium in Brno and by paying for their university education in the Medical Faculty of Vienna University. Alois was a medical student at the time of Mendel's last illness and he noted the symptoms of a chronic kidney condition which led to uraemia and dropsy. Recently an account of Mendel's last hours has come to light, which was written by one named Doupovec. It reads as follows:

> My mother often talked about Mendel and his last hours, for the duty of looking after Mendel was assigned to her and to a nun. She washed the bandages which were bound round the Abbot's feet and which needed changing many times a day. He suffered from loss of water, chiefly from the feet. It was a protracted and painful illness, but he rarely complained. He spent most of the time sitting on the sofa, only going to bed when he felt sleepy. The bandages were almost dry on the day of his death. My mother said to him: "Your Grace, today you have already no water." "Yes, it is already better," answered the Abbot. When the nun was making his bed she found him sitting on the sofa dead. . . .[26]

It was January 1884. The local newspaper *Brünner Tagesbote* paid the following tribute to him:

> His death deprives the poor of a benefactor, and mankind at large of a man of the noblest character, one who was a warm friend, a promoter of the natural sciences, and an exemplary priest. . . .[27]

NOTES

(1) Eichling, C. W. 1942. "I talked with Mendel." J. Hered. *33*, 245–246.
(2) Ibid., p. 244.

(3) Iltis, Mrs. Hugo. 1947. "Gregor Mendel's Autobiography." J. Hered. *38*, 234. (Referred to in future as "Au".)
(4) Ibid., p. 234.
(5) Iltis, H. 1932. *Life of Mendel translated by E. & C. Paul.* London. p. 42. (Referred to in future as "Il".)
(6) Il, pp. 66–67.
(7) Unger, F. J. A. N. 1853. *Botanical letters to a friend translated by B. Paul.* London. p. 107.
(8) Ibid., p. 18.
(9) Ibid., p. 16.
(10) Haberlandt, G. 1899. *Briefwechsel zwischen Franz Unger und Stephen Endlicher.* Berlin. p. 130. (Letter from Unger to Endlicher dated June, 1842.)
(11) Schleiden, M. J. 1849. *Principles of scientific botany; or, Botany as an inductive science. Translated by E. Lankaster from the 2nd German edition.* London. p. 457.
(12) MR1, p. 2.
(13) MR1, p. 2.
(14) Sajner, J. 1963. "Gregor Mendels Krankheit und Tod." Archiv Gesch. Med. *47*, 378. Also see: Křiženecký, I. J. 1963. "Mendels zweite erfolglose Lehramtsprüfung im Jahr 1856." Archiv Gesch. Med. *47*, 305–310.
(15) Fisher, Sir R. A. 1958. *The genetical theory of natural selection.* 2nd ed. New York. p. 9. (Dover edition of the work first published in 1929.)
(16) Fisher, Sir R. A. 1936. "Has Mendel's work been rediscovered?" Ann. Sci. *1*, 121.
(17) MR2, p. 3. (Letter from Mendel to Naegeli dated 18 April 1867.)
(18) MR2, p. 3.
(19) MR2, pp. 3–4.
(20) Il, p. 193.
(21) Mendel, G. 1911. "Über künstlicher Befruchtung gewonnenen *Hieracium*—Bastarde." Verh. naturf. Ver. Brünn, *49*, 53. (Reprint of 1869 paper.) Trans. in Il, p. 173.
(22) Bz, p. 429.
(23) MR2, p. 13. (Letter from Mendel to Naegeli dated 6 November 1867.)
(24) Prywer, C. 1932. "Genetische Studien über die Bastarde zwischen *Geum urbanum* L. und *Geum rivale* L." Acta Soc. Bot. Pol. *9*, 99.
(25) MR2, p. 15.
(26) Sajner, J. 1963. "Gregor Mendels Krankheit und Tod." Archiv Gesch. Med. *47*, 379.
(27) Il, p. 278.

SUGGESTIONS FOR FURTHER READING

Bennet, J. H. (Ed.) 1965. *Experiments in Plant Hybridisation. Mendel's original paper in English translation with Commentary and Assessment by the late Sir Ronald A. Fisher together with a reprint of W. Bateson's Biographical Notice of Mendel.* Oliver and Boyd, Edinburgh and London.

Correns, C. 1950. "Gregor Mendel's letters to Carl Naegeli. 1866–1873." Trans. L. K. and G. Piternick. Genetics, *35*, No. 5, pt. 2, 1–29.

Iltis, H. 1932. *Life of Mendel.* Trans. E. and C. Paul. Allen & Unwin, London; Hafner, New York.

Richter, P. "Johann Gregor Mendel wie er wirklich war. Neue Beiträge zur Biographie des berühmten Biologen aus Brünns Archiven." Verh. naturf. Ver. Brünn, *74*, pt. 2, 1–263.

Schindler, A. 1965. "Gedenkrede." (Privately printed in 1902.) Reprinted with a commentary by I. J. Křížencký in: Zaunich (Ed.) 1965. *Lebensdarstellungen deutscher Naturforscher.* Deutsche Akademie der Naturforscher Leopoldina in Halle. Halle.

Chapter Six

Pause and Rediscovery

The story of the almost simultaneous discovery of Mendel's classic paper and the rediscovery of Mendel's laws by three biologists in 1900 after thirty-four years of neglect is well known, but the circumstances and events which led to this discovery are little known. We will deal first with the facts of the story.

De Vries

The Dutch hybridist, Hugo de Vries (1848–1935) carried out experiments in hybridisation first in 1876. He crossed varieties of maize with sugary and starchy and black and white grains. Owing to the unfavourable weather conditions in Amsterdam in 1877 he obtained no harvest from the F1 hybrids.[1] This series of experiments was therefore terminated.

In the 1880s de Vries turned to the study of osmotic pressure in plant cells and made valuable contributions to the subjects of plasmolysis and the physiology of nastic movement.[2] Not until the close of the decade did he return to genetics and write his important book of 1889 entitled *Intracellular Pangenesis*, in which he stressed the need for treating the characters of organisms as separate entities in heredity and variation. Three years later he began once more to hybridise plants, starting with *Silene alba* × *S. alba* var. *glabra*, from which cross he obtained the following results in 1894:

536 F2 plants, 392 hairy and 144 smooth.[3]

This was his first numerical record of segregation but it does not appear to have led him to Mendel's explanation. In 1893 he crossed the poppies *Papaver somniferum* var. "mephisto", the flowers of which have a black heart-shaped spot at the base of each petal, with *P. somniferum* var. "Danebrog" which has white bases to the petals. Segregation took place in 1895 giving 158 black to 43 white. He self-pollinated these F2 plants and collected seed from 10 of the recessive white type hybrids and from 13 of the dominant black type. The following year, 1896, he sowed these seeds in separate plots and found that all the recessive segregates bred true (he admits finding two dominant F3 types in 1382), some of the dominant types bred true, and the others segregated once more giving 1,095 black to 358 white.[4]

Thus in 1896 de Vries had not only obtained a very good approximation to the 3 : 1 ratio (the theoretical expectation is 1,090 : 363) but also demonstrated the two-fold constitution of the dominant F2 segregates—part homozygous and part heterozygous.

Another cross, *Oenothera lamarckiana* × *O. brevistylis*, was also made in 1893 and showed segregation in 1895, "17 to 26% with the recessive character".[5] Unfortunately he gave no further details.

When H. F. Roberts wrote to him in 1924 asking how he came upon his discovery de Vries replied to the effect that it was the result of his "conception of unit characters" which he had stated in 1889 and his hybridisation experiments the result of which could be explained "as recombinations of these units".[6] The views he expressed in his *Intracellular Pangenesis* on hybridisation are well summed up in section 6, "Cross and Self-fertilisation", when he speaks of "how the individual hereditary qualities act as independent units in hybridisation experiments"; and in the final section of the book entitled *The Hypothesis of Intracellular Pangenesis* he describes these units as "more or less independent factors" which together contain in themselves potentially the character of the species. Evidently de Vries came to the experimental study of hybridisation as did Mendel with a factorial theory of inheritance already in mind. If we ask why he did not publish the results he had obtained up to 1896 in that year, the only possible answer is that he knew what Mendel knew in 1865, namely, success with a few species will not convince the world. In the autumn of 1899, however, he had obtained clear segregation with over thirty different species and varieties.

On Tuesday, July 11th, the Horticultural Society's International Conference on Hybridisation opened. In his introductory address Dr. Maxwell Masters regretted the absence of W. O. Focke, who was to have presided over the meeting, and the death of Charles Naudin. There followed five papers, the first by Bateson and the third by de Vries. The latter described his successful attempt to transfer the glabrous character of *Silene alba* var. *glabra* to *S. dioica*. His only reference to ratios was the following:

> . . . whilst in 1893 all the hybrids had been hairy, this was no longer the case in 1894. Only about three-fourths were hairy, the rest hairless. I had 99 hairy and 54 hairless, in all 153 plants, and counted them in July at the commencement of flowering.[7]

Now the theoretical expectancy here is 115 : 38. The frequencies he reported are much closer to a 2 : 1 ratio, so we can be certain that he had the much closer approximations to the 3 : 1 ratio in mind when he asserted that here was a case of $\frac{3}{4}$ hairy and $\frac{1}{4}$ glabrous. Why did he not take this grand opportunity to tell the world about his other experiments?—about his cross *Silene alba* × *S. alba* var. *glabra* which yielded 392 hairy to 144 smooth segregates in 1894? Was he timing it for the turn of the century?* He must have had good reason for remaining silent in 1899, and one can only conclude that he was

* See Note 39, p. 144.

still preparing his case, still collecting results for the great day which he had deferred so long.

We can imagine de Vries in the early part of 1900 flushed with the excitement of his discovery, confident after seven years of hybridisation experiments, preparing his brief notice for the *Comptes Rendus de l'Académie des Sciences* (*Paris*) entitled, "Sur la loi de disjonction des hybrides", and his longer preliminary report, "Das Spaltungsgesetz der Bastarde", for the *Berichte der deutschen botanischen Gesellschaft*. He finishes the French report and he is putting the finishing touches to the German report when there is a bolt from the blue. From his friend Professor Beijerinck in Delft he receives a reprint of Mendel's paper with the comment: "I know that you are studying hybrids, so perhaps the enclosed reprint of the year 1865 by a certain Mendel which I happen to possess is still of some interest to you."[8]

De Vries had been anticipated after all. How must he have felt? He had prepared the ground so well. First he established his priority in the field with his theory of intracellular pangenesis in 1889. Then he derived the law of segregation from the result of his experiments in 1895–1896. He withheld publication for four more years only to have his priority snatched from him by Mendel posthumously.

This reconstruction of the story, in my opinion, provides the most plausible explanation of why de Vries made no mention of Mendel in the French communication and why his references to Mendel in the German paper are very brief and appear to have been inserted into the text after the paper had been written. Anyone who takes the trouble to compare the two papers can see that the French paper is not, as T. J. Stomps would have us believe,[9] just a translation of that part of the German paper which contains a summary of results but no mention of Mendel. It seems much more likely that both papers were completed about the same time, and that de

Vries sent the French paper to a linguist with instructions to forward it to Paris after checking it. Meanwhile, Beijerinck's letter arrived, and all de Vries could do was to insert references to Mendel into the text of his German paper, redraft the conclusion and send it off post haste. Fortunately it arrived in Berlin before the French paper arrived in Paris; but as luck would have it, the French paper was published first. It contained not a word about Mendel.

Now why, one may well ask, did de Vries make no mention of Beijerinck's part in the discovery of Mendel's paper when he published his result in 1900? Why, when Roberts asked him about this discovery in 1924, did he reply that he first learnt of Mendel's paper from the bibliography given by L. H. Bailey in his book, *Plant Breeding* of 1895, and accordingly he "looked it up and studied it"?[10] Why did he tell the author of this book that it was from his article "Cross-Breeding and Hybridization" of 1892 "that I learnt some years afterwards of the existence of Mendel's papers, which now are coming to so high credit. Without your aid I fear I should not have found them at all"?[11] (Actually the citation of Mendel's papers occurs only in the 1892 essay and in editions of Bailey's book subsequent to 1895.)

In contrast to these remarks de Vries was much vaguer in 1900. Thus in a footnote to his German paper, "Das Spaltungsgesetz . . .", he remarked that he learnt of Mendel's paper "only after I had completed the majority of my experiments and had deduced the principles given in the text" (of this paper).[12] Had he really read Mendel's paper before Beijerinck sent him the reprint? Did he deliberately suppress reference to the paper, but changed his mind on receiving Beijerinck's communication? I think so. Otherwise it is difficult to see why he could write to Bailey and Roberts as he did, if he had not found Mendel's paper from Bailey's bibliography.

We will assume, therefore, that de Vries did find

Mendel's paper in this way, say in 1896 or 1897. His suppression of the fact may not have been so deliberate an act as it appears to us, for no doubt he underestimated the importance of Mendel's contribution compared with his own. I think he genuinely believed that his work was superior to Mendel's, that his own theoretical conceptions, expressed in 1889 and related to recent developments in cytology and to Weismann's theory of the germ plasm, put Mendel's ideas in the shade. Therefore there was no need to refer to him in the short two-page report which he sent to the French Académie, and no reason for giving more than passing reference to him in his more detailed German report. The most conspicuous change he made was in the conclusion:

French report:

> The totality of these experiments establishes the law of segregation of hybrids and confirms the principles that I have expressed concerning the specific characters considered as being distinct units.[13]

German report:

> From these and many other experiments I conclude that the law of segregation of hybrids in the plant kingdom, which Mendel established for peas, has a very general application and a fundamental significance for the study of the units out of which the specific characters are compounded.[14]

Had it not been for Mendel's paper which de Vries described as "trop beau pour son temps"[15] the law of segregation would have been known as de Vries' law. If that was what de Vries felt about it we can readily understand why he refused to add his signature to the Mendel Memorial Appeal in 1906. (See Appendix.)

Correns and Tschermak

For Correns and Tschermak the discovery that Mendel had anticipated them was not such a blow. In the first place they had in any case been anticipated by de Vries. Also, until 1899 the laws of heredity had not been the main object of their research. Correns had been studying sexual propagation in the foliose mosses and Tschermak the effects of in- and out-breeding. Correns had also been studying more complicated cases of inheritance which the discovery of Mendel's work helped him to elucidate. It accelerated the whole programme of his research, and he rated other items in this programme as greater achievements than his discovery of Mendel's laws, for he pointed out that as a result of all that had been discovered since 1865 "(I think above all by Weismann), the intellectual labour of finding out the laws anew for oneself was so lightened, that it stands far behind the work of Mendel."[16]

Tschermak said very little about his 3 : 1 ratios and gave no explanation of them. This was deliberate for he remarked to Roberts that: "The rules of inheritance, quite intentionally, I expressed at first purely descriptively or phenomenologically, in order not at once to anchor the newly-beginning experimental phase of the doctrine of heredity . . . to definite theoretical terms."[17]

We have seen that de Vries had already in 1896 results which approximated closely to the theoretical expectancy of 3 : 1. Correns and Tschermak did not reach this point until 1899. In the autumn of that year they arrived at the correct explanation. Each was unaware of the other's success. Then they discovered Mendel's paper as a result of reading the section on the *Leguminosae* in Focke's book *Die Pflanzenmischlinge* . . . 1881 in the winter of 1899–1900. When that winter drew to a close Correns and Tschermak were still pursuing their work unaware of any sense of urgency. Tschermak had written up his results as a doctoral thesis which he presented on January 17th, 1900. Then it went into the University archives under the control of the rectorate.

The only hint Correns had given of his success was in a veiled reference to ratios from maize hybridisations in the paper on xenia which he published in the *Berichte der deutschen botanischen Gesellschaft* at the end of 1899. From these hybridisations he obtained, he said, "very interesting but very complicated relationships".[18]

In April both men were greatly shaken to receive from Hugo de Vries a reprint of his brief report to the *Comptes Rendus*. Correns described what happened in the following words:

> On the morning of the 21st of April, 1900, I received a reprint "Sur la loi de disjonction des hybrides", of De Vries, and by the evening of the 22nd of April, my contribution, "G. Mendels Regel über das Verhalten der Nachkommenschaft der Bastarde", was ready. I sent it to the German Botanical Society, where it was received on April 24, and was reported in the session of April 27. The issue in question of the "Berichte" appeared at the end of May, about the 25th.[19]

Tschermak, likewise, lost no time in arranging for the publication of his work in the *Zeitschrift für das landwirtschaftliche Versuchswesen in Oesterreich* (Journal for agricultural research in Austria), reprints of the paper being available at the request of Tschermak in May, before the publication of the journal.

Hugo de Vries also sent a copy of his French report to William Bateson in Cambridge. Thereupon, Bateson hunted out Mendel's original paper—presumably the Royal Society copy, for the *Brünn Verhandlungen* was not available in Cambridge—and he was amazed by what he read. Mrs. Bateson tells us that he read it in the train on his way to a meeting of the Royal Horticultural Society where he was to lecture on "Problems of heredity as a subject of horticultural investigation". Presumably he was going to talk about de Vries' discovery so he had no difficulty in incorporating an account of Mendel's experiments. Hence it came about that on the afternoon of May 8th members of the Royal Horticultural Society heard about Mendel and his experiments which, said Bateson, "were carried out on a large scale, his account of them is excellent and complete, and his principles which he was able to deduce from them will play a conspicuous part in all future discussions of evolutionary problems".[20] Bateson became the apostle of Mendelism in England and he exploited to the full the promising field of experimental work which Mendelian genetics had opened up.

We have dealt now with the facts of the story. It remains to describe briefly the growth of ideas and knowledge which made possible the reception of Mendel's work in 1900.

The Analytic Approach

The most significant change of approach came when, instead of treating the species as a whole unit and the characters as expressions of that unit, the species was for the time being ignored and the characters were examined separately. The discoveries made by the cytologists and the programme of research set out by Darwin were largely responsible for this change. Thus in 1843 when cytology was still in its birth pangs Schleiden had grasped the importance of the analytical approach and had called for the recognition "as a fundamental principle in the study of the whole, the existence of an essential life in each separate cell".[21] Schleiden's campaign was continued by Rudolf Virchow who propounded the well-known aphorism "*Omnis cellula e cellula*", of which de Vries said in 1889:

> This proposition governs not only the Science of microscopy, but it is soaring ever higher to the mastery over the whole of biology. The fact that every cell has arisen from a material part of its mother cell, and that it owes its specific qualities to this origin, is true now for the doctrine of heredity as the basis of all thorough considerations.[22]

In 1859 Darwin sought to break down the species barrier so that he could establish the validity of applying to the origin of species his theory of the origin of varieties. Thereby he drew attention away from the species and directed it to the individual characters. Then in 1868 he put forward his hypothesis of pangenesis which differed from other theories of particulate inheritance, such as those of Herbert Spencer and Naegeli, by the fact that his individual genetic particles determined not the whole organism but each a separate part of the whole. The idea of separate particles for the various characters was, according to de Vries, responsible for his concept of "pangenes", and he regarded the shaking up of the old species concept as one of the most important benefits from Darwinism. "If", said de Vries, "one considers the specific characters in the light of the theory of descent, it soon becomes clear that they are composed of individual more or less independent factors."[23] And every species "when looked at in this way, appears to us as an extremely complicated image, but the entire organic world is the result of innumerable and diverse combinations and permutations of relatively few factors".[24]

Bateson

William Bateson was led to the same idea from his study of variation. Starting as an orthodox Darwinian he had searched for a causal connection between variation and environment. For this purpose he set out for Turkestan in 1886 to study the fauna of the lake basins which were gradually drying up. Having failed to find what he was seeking he turned to the study of records of variation in man and domesticated animals. His results published in *Materials for the Study of Variation* 1894, convinced him of the discontinuity of variations, a characteristic which, he said, "is in some unknown way a part of their nature, and is not directly dependent upon Natural Selection at all".[25] Hence for him evolution by continuous variation was "a gratuitous assumption" and the swamping effect of crossing a "far-reaching mischievous error". To get at the truth, he declared, we must organise "... experiments in breeding, a class of research which calls perhaps for more patience and more resources than any other form of biological inquiry. Sooner or later such investigations will be undertaken and then we shall begin to know".[26]

A year later Bateson was off to Italy where in the Val Formazza he found the butterfly *Pieris rapi* var. *bryoniae* which he later crossed with the variety *egeria*. Also in the Val Formazza he noted a glabrous and a hairy variety of the Cruciferous Spectacle plant (so named on account of the appearance of its fruits), *Biscutella levigata*, growing together. His colleague Miss Saunders, a Fellow of Newnham College, Cambridge, raised plants of these two varieties in an allotment behind the University Botanic Garden in August 1895. Crosses were made in 1896, but as no F2 plants were raised Miss Saunders made no mention of ratios in the report she submitted to the Royal Society in 1897. That year, however, a fresh start was made using more promising material: *Datura stramonium* × *D. stramonium* var. *inermis*, both of which involved the single character difference prickly to smooth fruits; *Silene alba* × *S. alba* var. *glabra*, hairy to glabrous leaves; and varieties of biennial and annual stocks, *Matthiola incana*, which differed in character of leaves, hoary or glabrous, colour of seeds and flowers and time of flowering.

Bateson's butterflies became diseased and his stocks (*Matthiola*) were killed off by the severe winters so that in 1899 segregation was seen only in:

Datura	prickly to smooth fruits	F2	152 : 43
		(expected	146 : 49)
	prickly to smooth fruits and red to white flowers + red to green stems	F2	204 : 65 : 81 : 13
		(expected	204 : 68 : 68 : 23)

Silene	hairy to glabrous leaves	F2	408 : 126	
		(expected	401 : 133)	
		backcross	447 : 433	
		(expected	440 : 440)	27

The frequencies in *Datura* give scarcely any more indications of ratios than do those of Naudin's obtained with *Datura* half a century earlier, but in the case of *Silene* there is an indication. Even so, the truth did not dawn on Bateson in 1899 although his experimental approach was by that time so like Mendel's that he talked just like a Mendelian. This is illustrated by his paper "Hybridisation and cross-breeding as a method of scientific investigation" which he read at the International Conference on Hybridisation, 1899, on the same day that de Vries read his paper. Speaking of the need for facts about the evolution of a particular form he said:

> What we first require is to know what happens when a variety is crossed with its *nearest allies*. If the result is to have a scientific value, it is almost absolutely necessary that the offspring of such crossing should then be examined *statistically*. It must be recorded how many of the offspring resembled each parent and how many showed characters intermediate between those of the parents. If the parents differ in several characters, the offspring must be examined statistically and marshalled as it is called, in respect of each of those characters separately. Even very rough statistics may be of value. If it can only be noticed that the offspring came, say, half like one parent and half like the other, or that the whole showed a mixture of parental characters, a few brief notes of this kind may be a most useful guide to the student of evolution.[28]

From this statement it seems as if he did realise that the F2 generation segregates according to simple ratios, but if he did, the reason was not apparent to him. Hence it came about that when he read de Vries' paper a year later the explanation it contained was new to him.

Cytological Basis of Mendel's Laws

It has been suggested that even if the hybridists had not discovered Mendel's laws they would have been predicted by the cytologists; for when they unravelled the life history of the chromosomes in the 1880s and 1890s they discovered the processes and particles upon which Mendelian inheritance depends.

The most important stages in the unravelling of this story were first the discovery of the longitudinal division of the chromosomes by which successive daughter chromosomes maintain their identity throughout the organism and successive daughter nuclei receive the same number of chromosomes. Then came the discovery that there are two sets of chromosomes in all the body cells of the organism but only one set in the germ cells. This was material evidence for Mendel's implied inference of the double representation of hereditary factors in the body cells and their single representation in the germ cells. When the details of the reduction division or meiosis which causes this halving of the chromosome number were elucidated it was seen that the distribution of the members of each chromosome pair to the germ cells is equal but it is a matter of chance as to which member goes into which germ cell. Providing therefore large numbers of offspring are produced, the various combinations of characters will all be realised equally. Hence the numerical ratios demonstrated by Mendel are the result of the random movement of chromosomes into sister germ cells and the random fertilisation between germ cells.

The relationship between the mechanics of meiosis and the laws of inheritance was not fully understood until long after 1900 but already in that year Correns had tentatively identified germinal segregation with the first division of the pollen and embryo mother cells[29] (in fact it can be the first or the second, depending on where crossing over takes place).

When Mendel read his paper in 1865 none of these cytological discoveries was available in support of his case. This is why the situation thirty-five years later was so different. In that interval of time not only had important discoveries been made but August Weismann had also prepared the way. He made biologists conscious of the importance of the chromosomes by identifying the hitherto hypothetical genetic particles with them. He attached great importance to the reduction division and he realised that here was a source of variation, for if one set of genetic elements (he called them "ancestral germ-plasms") were expelled from one egg during meiosis and another set from another egg "it follows that no two eggs can be exactly alike as regards their contained hereditary tendencies".[30] Hence "the germ cells of any individual do not contain the same hereditary tendencies, but are all different, in that no two of them contain exactly the same combinations of hereditary tendencies. On this fact the well-known differences between the children of the same parent depend."[31] This, we have seen, was precisely the point which Naegeli and Darwin were not prepared to accept. Weismann's innovation prepared biologists for the reception of Mendelian heredity and his reflections on the discoveries of the cytologists prepared both parties—cytologists and hybridists—for the forthcoming union of their studies.

Mendel's Law

Finally we must ask ourselves the question: What are Mendel's laws and to what extent are they still applicable today? Some writers are under the impression that Mendel did not himself frame any laws. This just is not so. At least seven times Mendel refers to "das Gesetz"—"the law governing *Pisum*"; "the law which is valid for *Pisum*", "the law of development discovered for

Pisum".[32] He called it a law because it applied to all the characters he had investigated in *Pisum*, some of the characters in *Phaseolus*, and he also believed that it would apply to all other plants. Until experimental proof for this is forthcoming he said, "we may assume that in material points an essential difference can scarcely occur, since the unity in the developmental plan of organic life is beyond question".[33]

Mendel's law was the "Law of combination of different characters", i.e. what we now term the independent assortment or recombination of characters. When he stated this law in the earlier part of his paper he kept very close to the facts and made no definite statement about segregation. The latter is of course implied in the following statement:

> . . . in the ovaries of the hybrids there are formed as many sorts of egg cells, and in the anthers as many sorts of pollen cells, as there are possible constant combination forms . . .[34]

But it was not until he had described all his results and had formulated the law of the independent assortment of characters that he went on to discuss how a hybrid can form pure germ cells. Then he suggested that there is a temporary association of conflicting elements in the F1 hybrids which is broken when the germ cells are formed. He pictured this process as a mutual separation of the elements which determine each pair of differentiating characters, but an entirely free association of the elements responsible for different traits. In 1865 one could hardly have expected a more accurate prophecy of the mechanism of segregation—the separation of homologous chromosomes in meiosis. The passage containing this important statement is reprinted in the appendix to this chapter.

In 1900 Correns redefined Mendel's law as follows:

> In the hybrid reproductive cells are produced in which the *Anlagen** for the individual parental characteristics are contained in all possible combinations, but both *Anlagen* for the same pair of characters are never combined. Each combination occurs with approximately the same frequency.[35]

And he added that it includes de Vries' law of segregation. Though Correns retained the designation "Mendel's *Law*", it became the custom to speak of "Mendel's *Laws*". Segregation became the first law and independent assortment the second. This procedure is perfectly justified providing we remember that Mendel himself did not adopt this practice.

Whilst we can claim that the fact of the purity of the germ cells was stated by Mendel and a mechanism of germinal segregation was suggested by him, we cannot also attribute to him the assertion of a finite number of hereditary elements per character-pair. There is indeed much in his paper which hints at two hereditary elements per character-pair as the simplest case, and multiples of two for more complicated cases. Thus he deliberately dealt with qualities which have two quite distinct expressions, and he treated these opposing expressions in pairs calling them the character-differences. The step from representing any one of these pairs by *two* letters to the assertion that the qualities in question are each determined by two elements is surely a natural one to take. And when he represented hybrid forms by two letters (*Aa*) and germ cells by single letters (*A* and *a*) he must have felt the temptation to represent constant forms by two letters as well (*AA* and *aa*), but he did not, for in his paper these letters represent simply the classes to which the plants and the germ cells belong.[36] It is as if he deliberately avoided drawing the conclusion which his discovery of the 1 : 2 : 1 ratio so strongly implied. If, on

* The German word "Anlage" is difficult to translate but elements or factors are possible equivalents.

the other hand, the process of reduction division had been discovered in the 1860s instead of in the 1880s Mendel would surely have taken the plunge.

Exceptions to Mendel's Law

Mendel recognised one exception to his law, and one modification of it. The former was in *Hieracium* where he realised that a "permanent union" of germinal elements may occur. The latter was in the flower colour of *Phaseolus* where inheritance is made more complex than in *Pisum* by the interaction of factors. Genic interaction has now been shown to occur widely and it modifies the numerical ratios now obtained. Other conditions which limit the validity of Mendel's law are: polyploidy, reproduction in triploid hybrids does not conform to the Mendelian situation since survival of the various germ cells is low and differential, i.e. they are not produced in equal numbers. Therefore, the numerical ratios in the F2 generation do not conform to Mendel's law. Linkage, or "gametic coupling" as its discoverer Bateson called it, prevents the independent segregation of factors which are on the same chromosome. Mendel's assertion that the constant characters "may be obtained in all the associations which are possible according to the laws of combination"[36] can apply only to factors on different chromosomes. Thus where Mendel would have obtained a 9 : 3 : 3 : 1 ratio for two pairs of contrasted characters only a 3 : 1 ratio is obtained if the factors are linked. This inhibition of independent association is mitigated somewhat by crossing over. With crossing over one might for instance obtain 11 : 1 : 1 : 3.

To the question does Mendel's law still hold, the answer is yes; but, like any other scientific law, it holds only under prescribed conditions. Mendel stated most of these conditions, but the need for no linkage, crossing-over and polyploidy were stated after 1900. In addition, cases of non-Mendelian inheritance were discovered as early as 1909 by Carl Correns and Erwin Baur, which have since proved to be due to determinants outside the nucleus. Mendel's law applies only to determinants on the chromosomes.

De Vries stated two principles in 1900:

1. The hybrid always shows only one of the two antagonistic qualities.
2. The two antagonistic qualities separate in the formation of the pollen and egg cells.

The second principle is the law of segregation. The first is de Vries' law of dominance, which both Correns and Bateson criticised, which was not stated by Mendel and today is not recognised as a law at all. Mendel was not without his critics in 1900 and they naturally exploited such superficial conclusions as this law of dominance in their attempt to invalidate all Mendel's work. Now that the battle for Mendelism is over, Mendel's own paper remains, as it ever will do, a great classic in the history of science; a record of a powerful mind probing as had Koelreuter a century before, into one of nature's most guarded secrets; of a writer shaking himself free from the vague terminology and imprecise language of his predecessor Gaertner as he strove to expound lucidly the new theory he had formulated.

Conclusion

We may regard Koelreuter's work of two centuries ago as the foundation stone of experimental genetics. We can see the language of genetics begin to take its shape for the first time in Gaertner's book of 1849, after which it was made more precise by Mendel in 1865 and enriched and modernised by his successors at the beginning of this century with fresh terms coined by Bateson and extant terms borrowed from the cytologists. The contemporary theory of genetics, however, is still justly termed "Mendelian" since it is Mendel's principles and not those of his predecessors which are still valid today as an explanation of the nature of heredity.

It has become customary to look upon the neglect of Mendel's work for thirty-four years as one of the most singular events in the history of science, but as we have seen there were good reasons for its neglect. Mendel was a sensitive and a humble man whose contact with scientists was limited to central Europe. Although he visited England he never met British scientists, and most important of all—he never met Francis Galton. The history of genetics would surely have been different had these two original thinkers met. Mendel published his work locally because he had lectured on it locally, but this does not mean that he would have refused to publish it in the *Berichte der deutschen botanischen Gesellschaft* if he had been requested to do so. But he did not dare to ask. So he wrote to Naegeli and Kerner von Marilaun first, hoping for encouragement; but he did not get it. They were far too busy studying highly variable genera which showed promise of evolving in front of their eyes, to stop and examine the laws of inheritance which everyone assumed were expressed by the fractional law $\frac{1}{4}p + \frac{1}{8}pp + \frac{1}{16} \ldots$ The neglect was thus in part a reaction between personalities, in part a result of assuming that the study of constant forms will not lead to the source of variation, and in part to the simple fact that the cytologists had yet to discover the material basis for Mendelism and for a concept of heredity far more non-variable than any Darwinian was prepared to accept.

NOTES

(1) De Vries, H. 1900. "Sur la fécondation hybride de l'endosperme chez le mais." Rev. gén. Bot., *12*, 134.

(2) Stiles, W. 1950. *An introduction to the principles of plant physiology*. 2nd ed. London.

(3) De Vries. 1903. *Die Mutationstheorie. Versuche und Beobachtungen über die Entstehung von Arten im Pflanzenreich*. Vol. 2, Leipzig. p. 157.

(4) Ibid., p. 165.

(5) Ibid., p. 157.
(6) Roberts, H. F. 1929. *Plant hybridisation before Mendel.* Princeton. p. 323. (Referred to in future as "Roberts 1929")
(7) De Vries. 1900. "Hybridising Monstrosities." J.R. Hort. Soc., *24*, 74.
(8) Stomps, T. J. 1954. "On the rediscovery of Mendel's work by Hugo de Vries." J. Hered., *45*, 294.
(9) Ibid., p. 294.
(10) Roberts, H. F. 1929. p. 323.
(11) Bailey, L. H. 1908. *Plant Breeding.* 4th ed. p. 155.
(12) De Vries. 1900. "Das Spaltungsgesetz der Bastarde. Vorläufige Mitteilung." Ber. dtsch. bot. Ges., *18*, 85.
(13) De Vries. 1900. "Sur la loi de disjonction des hybrides." C. R. Acad. Sci., Paris, *130*: Trans. in: "The birth of Genetics Mendel–de Vries–Correns–Tschermak in English Translation." Genetics Supplement 1950, *35*, No. 5, pt. 2, p. 32. (Referred to in future as MR3.)
(14) De Vries. 1900. "Das Spaltungsgesetz der Bastarde. Vorläufige Mitteilung." Ber. dtsch. bot. Ges., *18*, 90.
(15) De Vries. 1900. "Sur les unités des caractères spécifiques et leur application à l'étude des hybrides." Rev. gén. Bot., *12*, 271.
(16) Roberts, 1929. p. 337.
(17) Ibid., p. 345.
(18) Correns, C. 1899. "Untersuchungen über die Xenien bei *Zea Mays*." Ber. dtsch. bot. Ges., *17*, 411.
(19) Roberts, 1929. p. 337.
(20) Bateson, W. 1900. "Problems of heredity as a subject for horticultural investigation." J. R. hort. Soc., *25*, 59.
(21) Schleiden, M. J. 1849. *Principles of scientific botany; or, Botany as an inductive science.* Translated by E. Lankaster from the 2nd German edition. London. p. 457.
(22) De Vries. 1889. *Intracellulare Pangenesis.* Jena. pp. 75–76.
(23) Ibid., p. 7.
(24) Ibid., p. 9.
(25) Bateson. 1894. *Materials for the study of variation treated with special regard to discontinuity in the origin of species.* London. p. 573.
(26) Ibid., p. 574.
(27) Bateson & Miss Saunders. 1902. Reports to the Evolution Committee of the Royal Society: Report 1. *Silene*—p. 17, Table II, p. 18, Table III. *Datura*—p. 25, Table VII, p. 26, Table VIII.
(28) Bateson. 1900. "Hybridisation and cross-breeding as a method of scientific investigation." J. R. hort. Soc., *24*, 63. (1899 Hybrid Conference.)
(29) Correns, C. 1900. "G. Mendel's Regel über das Verhalten der Nachkommenschaft der Rassenbastarde." Ber. dtsch. bot. Ges., *18*, 164, footnote 1. (For a lucid explanation of this point see: Lewis, K. R. and John, B. 1964. *The Matter of Mendelian Heredity.* Churchill, London. p. 49.)
(30) Weismann, A. 1889. *Essays upon heredity and kindred biological Problems.* Oxford. Essay VI, "On the number of polar bodies and their significance in heredity 1887." p. 379.
(31) Ibid., p. 384.

(32) MR1, pp. 30, 34, 28.
(33) MR1, p. 37.
(34) MR1, p. 20.
(35) MR3, p. 40.
(36) Bennet, A. J. 1964. "Mendel's laws?" Sch. Sci. Rev., 46, No. 158, 37.
(37) MR1, p. 19.
(38) De Vries. 1900. "Das Spaltungsgesetz der Bastarde. Vorläufige Mitteilung." Ber. dtsch. bot. Ges., *18*, 84.
(39) In the C.R. Acad. Sci., Paris, for 4. XII. 1899, De Vries reported the numbers 180 and 66 and the fractional segregation of about ¼ and ¾, for endosperm colour in maize obtained in 1898 and 1899 respectively, but no explanation or reference to Mendel.

SUGGESTIONS FOR FURTHER READING

Darlington, C. D. 1964. *Genetics and Man.* (Revised edition of *The Facts of Life.* 1953.) George Allen & Unwin, London; Hillary, New York.

Jahn, I. 1957–58. "Zur Geschichte der Wiederentdeckung der Mendelschen Gesetze." Wissenschaftliche Zeitschrift der Friedrich-Schiller Universität Jena, mathematisch-naturwissenschaftliche Reihe, Jahrgang 7, pt. 2/3, 215–227.

Krumbiegel, I. 1957. *Gregor Mendel und das Schicksal seiner Vererbungsgesetze.* (Grosse Naturforscher, Bd. 22.) Wissenschaftliche Verlagsgesellschaft M.B.H., Stuttgart.

Tschermak-Seysenegg, E. von. 1951. "The rediscovery of Gregor Mendel's work". J. Hered., 42, 163–171.

Zirkle, C. 1964. "Some oddities in the delayed discovery of Mendelism." J. Hered., *55*, No. 2, 65–72.

Notes on the Appendices

Extensive quotations have not been made in the main text, but in the appendices the reader will find the more important passages cited in full. The majority of these have been translated from the German, French or Latin originals. Where a direct translation would have produced a clumsy end product strict accuracy has been sacrificed and a free translation provided.

Appendix—Contents

APPENDIX TO CHAPTER 1

Linnaeus' Hybrids

Among plants, an outstanding example of a hybrid is afforded here at Uppsala by a plant originating from *Veronica maritima* fertilised by *Verbena officinalis*; in this hybrid offspring it is still possible to see female productive organs, derived from the Medulla, which closely resemble the mother; but external features, such as the leaves and other cortical parts, which strongly resemble the father. Several years ago, in a bed of the University Botanic Gardens, where *Tragopogon pratensis* and *Tragopogon porrifolium* were growing together, a plant arose which recalled *Tragopogon pratensis* but had reddish flowers; so, as an experiment our President in the year 1757, took a plant of *Tragopogon pratensis* which had just opened its flowers, blew strongly on them to disperse the pollen and lightly sprinkled them with genital dust (pollen) from a flower of *Tragopogon porrifolium* which he had plucked. The experiment was repeated for several days running on the same flower; eventually the flower set seed; seeds were sown in 1758, and this year (1759) have flowered and fruited. It was noteworthy that its flowers or corollae were purple almost to the base; that its calyx was longer than in *pratensis*, and the peduncles rather thicker than usual; in a word, it resembled the father in outward features, the mother in its inner parts.

Ramstrom, C. L. 1763 "Generatio Ambigena, quam praeside D. D. Car. Linnaeo, . . . Upsaliae 1759 . . ." Amoenitates Academicae, *6*, 11–12.

Buffon: a Hybrid between Wolf and Dog

Buffon's attempt to hybridise these two failed because

the wolf killed the dog, but M. Surirey de Boissy wrote the following letter to Buffon describing a successful attempt.

Namur, 9th June, 1773. At M. le Marquis de Spontin's at Namur a very young she-wolf was reared, who for two years had as a companion a dog of about the same age. They were at liberty to wander through the house, kitchen and stables and they slept under the table or on the feet of those around them. They lived on very intimate terms.

The dog was a very vigorous breed of mastif. The wolf's diet for the first six months was milk, after which they gave her raw meat which she preferred to cooked meat. While she was eating no one dared to approach her; at other times you could do what you liked with her as long as you did not maltreat her. She would caress all the dogs which were brought to her until the time when she showed a preference for her old companion, then she would quarrel with all others. She was first mated on the 25th March [1773]. The mating was repeated frequently over a period of sixteen days. At eight o'clock on the sixth of June she gave birth to her young ones and so the gestation period was seventy-three days at the most. She gave birth to four puppies. They have the extremities of their paws and one half of the chest white, in this holding to the dog, which is black and white. Since being brought to bed she has growled and bristled at anyone who approached. She would not recognize even her masters and would even fly at the dog if he came near.

I add that she has been double-chained (*attachée à deux chaines*) after an outburst that she made chasing the dog who had leapt over a neighbour's wall where there was a bitch on heat. She half-strangled her rival and the coachman used a stick to separate them. He conducted her back to her quarters where, owing to an unwise repetition of the correction [beating her with a stick] she was enraged to the point of biting him twice in the thigh, as a result of which considerable incisions had to be made which kept him in bed for six weeks.

Buffon. *Histoire Naturelle* . . . Supplément 3, 1776, pp. 9–10.

Koelreuter: his First Hybrid Plant

The plants which some of the recent teachers of botany have boldly declared to be hybrids may well be no other than untimely births of an exaggerated imagination. Possibly there are a few which with justice may deserve this name. How can one with certainty make a statement on the matter before one has produced them by art and indeed by the most reliable experiments? It is just as unlikely that a hybrid plant should arise in the orderly arrangement which nature has made in the plant kingdom as that a hybrid should have arisen at any time from two kinds of animals living in their natural freedom. Nature which, even in the greatest apparent disorder, ever observes the most wonderful order, has prevented this confusion, in the case of wandering animals, besides other means, chiefly by the natural instincts, and in the case of plants, where their much too close proximity, the wind and insects daily give the opportunity for an unnatural union, she will have known without doubt by just as certain means how to take away the force of the alarming consequences which they present. Presumably they are the same as those means, other than natural instincts, which are found in animals. It is also possible that, to prevent such an alarming disorder, one of her designs has been to transfer one plant to Africa and assign to another its place in America. Possibly it is partly for this reason that in the limits of a certain region only such plants are included as have the least similarity in respect of structure and which consequently are the least suited to produce disorder amongst one another.

If these surmises have some foundation, as I almost believe, then hybrid plants will probably be able to arise in botanic gardens, where plants of all kinds and from all parts of the world are together in a restricted space, particularly if one puts them together according to a systematic arrangement . . . Here, at least, man gives plants in a positive manner just the opportunity which he gives to his animals, often brought from widely-spaced parts of the earth, which he confines contrary to nature in a zoo or in a yet more restricted space. I wonder whether a goldfinch would ever have mated with a canary and produced hybrid offspring if man had not provided them with the opportunity of coming to know one another more closely? Ought not hybrid plants already to have arisen in botanic gardens? The very grounds which make me doubt the production of the same under natural conditions move me to admit it under unnatural conditions.

Because I was already long ago convinced of the sexuality of plants and at no time doubted the possibility of such an unnatural production I allowed nothing to deter me from planning experiments thereon, in the good hope that possibly one day I might be so fortunate as to bring about the production of a hybrid plant. After many attempts made with many kinds of plants I have at last in the previous year, 1760, progressed so far in the case of two different species of a natural genus, namely in the case of *Nicotiana* (*paniculata*) Linn. Sp. Pl. p. 180. n. 2 and *Nicotiana* (*rustica*) Linn. Sp. Pl. p. 180. n. 3., that I have fertilised the ovary of the former with the pollen of the latter, obtained perfect seeds and from these have raised young plants, all in the same year. Since I have made this experiment with many flowers, at various times and with all possible precautions, and have every time obtained an ordinary fertilisation and perfect seeds, therefore I could not for a moment believe that by chance there had been a mistake in the experiment, and that the plants already arisen from the seeds, of which seventy-eight had come from 110 seeds, might be just ordinary mother plants. Although I could not at first discover much in them that was especial and unusual, yet I had already found a noticeable distinction between the seeds produced naturally and those produced artificially which gave me less cause for doubting the hybrid nature of the plants produced from them. Finally I was convinced completely when a little over twenty of them which I had kept through the winter, some in rooms and some in a cold greenhouse, came into flower last March. I became aware, with much pleasure, that they held exactly to the mean between the two natural species not only in the spread of the branches and in the position and colour of the flowers, but also especially in the parts belonging to the flower save only the stamens which showed an almost exact geometrical proportion between the two natural species. This circumstance perfectly justifies the ancient Aristotelian doctrine of procreation by means of both kinds of seed, and entirely contradicts the doctrine of seminal animals,* or of the original seed assumed to be in the ovary of animals and plants which gives rise to living embryos and germs under the influence of the male seed.†

The anthers were noticeably smaller than those produced from both the natural plants, and consequently they contained less seed dust (pollen) in their interior than did the latter. It was also paler and drier and its parts did not so firmly stick together. This peculiar circumstance led me immediately to examine the same with a magnifying glass. Just as all the other parts of the hybrid were perfect so this part was imperfect. For, having established that the pollen grains of both natural species have a regular elliptical shape and are full of male seed, so these were quite irregularly shrunken and, as it were, pulverised; they contained scarcely any fluid material and were,

* The doctrine of the spermists.

† The doctrine of the ovists.

in a word, mere empty husks. Hence the fertility of this plant appeared to me at once to be extremely suspicious and the result justified my suspicion completely, for amongst an innumerable quantity of flowers not one was found which had born even a few seeds although they were covered with a large quantity of their own pollen . . . OK, pp. 29–31.

Koelreuter: F2 Hybrids

First Continuation

VII Experiment

	rustica ♀
Nicotiana ☿	
	paniculata ♂

propr. pulv. consp.

Amongst the infertile hybrids which were mentioned in my *Preliminary Report* p. 30 were some which indeed agreed in external aspect with the rest but which showed a very slight degree of fertility on the male side, and from their own pollen gave occasionally one or a few small seeds. . . . I covered one of these fertile hybrids with its own pollen with the greatest care, and from the seeds resulting therefrom I have raised plants which were no ordinary hybrids but such as have been reported in paragraphs 3 and 5.* I allowed four of them to grow to completion. Their pollen consisted of a so great quantity of good perfect pollen grains that fertile capsules often containing 200 good seeds resulted from most of the flowers. Since one sees clearly from this how vigorously such a small quantity of good pollen, which the above fertile

* The hybrids described in paragraphs 3 and 5 were the back-crosses: *Nicotiana rustica* × *paniculata* × *rustica* which varied but were all more like *rustica*, and *Nicotiana rustica* × *paniculata* × *paniculata* which were all more like *paniculata*, some to a greater degree, some to a lesser.

hybrids yielded, is increased at once in the next generation, it is thus highly probable that if one were to cover these plants repeatedly with their own pollen, in time and possibly in a few years, they would be transformed into perfect mother plants. From this I draw the following conclusion:

That imperfect hybrids which possess a slight degree of fertility on the male side appear to change back to mother plants again by their own forces.

OK, pp. 52–53.

Koelreuter: Plant Reproduction

For the procreation of every natural plant two homogeneous fluid materials are required which differ in nature and are intended for union with each other by the Creator of all things. The one thereof is the male seed, the other the female seed. Since these materials differ in nature, . . . it is easy to understand that the force of the one must be different from the force of the other. From the union and commingling of these two materials, which takes place in the most intimate and orderly manner in definite proportions (*nach einem bestimmten Verhältnisse*), another material arises which is of an intermediate nature and hence possesses a compound force derived from the two former simple forces, just as in the union of an acid salt and an alkaline salt, a third, namely a neutral salt,* is formed. Following the occurrence of commingling, this third material either at once constitutes the beginning or the firm foundation of a living machine, or it brings it forth in a little while. . . . Never would one of the seed materials alone have been able to produce anything similar: for no more can a pure acid salt or a pure alkaline salt alone

* Acids and bases used to be known as salts, hence the terms "acid salt, alkaline salt and neutral salt". In the latter part of this passage Koelreuter expresses the process of germinal segregation in terms of his fluid theory of fertilisation.

become a neutral salt, and itself form a crystal. The gradual formation of the future plant, its particular organic structure or its specific nature by which it is distinguished from all others . . . rests upon this foundation and its operative force which must necessarily be different according to the different nature of its twofold seed material in every particular kind of living machine. . . . All movements and changes which take place from germination to the time of flowering in all such masterpieces of nature, appear to be aimed solely at the great work of reproduction at which they work as it were with united forces. They all aim at liberating bit by bit the one compound material from which they are formed and dividing it again into the two original basic materials, or more correctly, to yield the latter in whole masses which are unequal in size especially from the one side, as was shown in the preceding reproduction.

OK, pp. 42–43.

APPENDIX TO CHAPTER 2

Henschel: Opposition to the Sexuality of Plants

Once one has entered the magic circle, whose centre is the sexual plant, it seems that it is almost impossible to leave it, for an intuitive knowledge (*Erkenntniss*) which is indisputably proven, illuminates it from its innermost centre. The plant, it is said, must have a procreation, a sex. If it lives it must have been procreated, since procreation and life are one. Procreation is only the overflowing of life; life is only a procreation flowing back into itself. Procreation cannot take place without the sexes, therefore the plant must also possess the procreating sexes, and it cannot exist and live without them.

Around this central intuitive knowledge (*Einsicht*) the

truth of which has been accepted consciously or unconsciously, and is still accepted today, a mound of facts based on years of experience and observation has accumulated. It surrounds the sacred centre like an impenetrable rampart. Every empirical botanist is attracted into it and no matter how forceful and able he may be, he cannot escape from it. Here it seems that for the first time empirical science has hammered out the truth single-handed and has secured it safely for all time.

Nevertheless, the voice of opposition against this doctrine has never quite been silenced. . . . Even very recently . . . the ancient dispute has been stirred up again by F. J. Schelver. Now why is it that this sex of plants, after being studied for two centuries, is still being attacked by presumptious opponents?

Henschel, A. 1820. *Von der Sexualität der Pflanzen*. Breslau. pp. iii–iv.

Henschel: on the Infertility of Koelreuter's Hybrids

Have plant hybrids the essential qualities of animal hybrids? Infertility with its own kind and with other species is reckoned as the principal quality of animal hybrids. Only Koelreuter speaks of hybrids which did not entirely possess this quality. Some fruit formation took place on hybrids which had been dusted with paternal or maternal pollen or with their own pollen. (p. 447) . . .

The hybrid *Dianthus chinensis* × *D. carthusianorum* . . . gave a number of good seeds after self-pollination. The hybrids between *D. chinensis* and a sweet william, which Gmelin brought from Siberia, were even more remarkable in this respect; Koelreuter says of them: "It is very remarkable that these new hybrid pinks had a fairly high degree of fertility on both sides. For not only did plants growing in the open produce a quantity of fertile fruits, but those which I had dusted with their own pollen or with that of the Chinese and the Carthusian pink gave usually 20 to 30 fairly large, black, perfect seeds. Indeed, when several Chinese pinks were dusted with the pollen of these hybrids, the seed-vesicles in the ovary were nearly all perfectly fertilised." (p. 448) . . .

Very many quite distinct species formed fertile hybrids with each other. In these cases Koelreuter made a strange *a posteriori* conclusion; he supposed that because they gave rise to fertile individuals one should not regard the differences of the parents as specific in nature but merely as varietal. (pp. 448–449) . . .

Infertility was mainly the result of two kinds of conditions. (1) The results of external conditions. It is known that carefully cultivated plants are generally infertile even though the greater care, the fertile soil and the precise regulation of external influences enhances the individual life more than the general life. Consequently it appears that the research plants which were raised in pots, crammed into a narrow space, and thereby restricted in their growth, seemed more fertile than those growing in the open ground, e.g., the *Verbascum* species. That this effect of the conditions of life has operated throughout Koelreuter's experiments and hence that careful culture may have been the cause of infertility in the majority of cases is the more likely since he cites the following characteristics as common to all hybrids: more rapid growth, accelerated and extended flowering period, additional sprouting from stem and root in the autumn, indeed even longer life of the whole plant, the effect of which is to increase perceptibly the individual vegetative force and to antagonise and suppress the general force. Moreover, the season and the same mysterious ordering power of nature (*Naturordnung*), which allows one plant to become fertile and another infertile, acts on the individual plants and has quite by chance made this one infertile whilst in the next year it has served it better. Even Koelreuter found that *Verbascum phoeniceum* growing in various situations in the wild remained infertile for three successive seasons. Why, then, ought one to consider the infertility of a variant arising from this species as a symptom of a hybrid constitution?

The second important circumstance, which had an effect on the fertility or sterility of the hybrids, lay often in the nature of the plants taking part in the hybridisation. It is still remarkable that the hybrid *Dianthus plumarius sib.* × *D. chinensis* was very fertile, that *D. plumarius* × *D. glaucus* was sterile, that the hybrid *Nicotiana rustica* × *N. paniculata* pollinated by *N. rustica* was only partially sterile and when it was pollinated by *N. paniculata* it appeared quite sterile. Hence it is clear that the degree of fertility is related to the relationships between the inner nature of the species involved.

Henschel. *Von der Sexualität der Pflanzen*. 1812. pp. 450–452.

Gaertner: on Evolution

Hybrid fertilisation is still considered as the plan and purpose of nature by many botanists, especially by those who believe that genera which comprise many species can only have arisen by hybridisation. . . . Koelreuter has already disputed this hypothesis, and it will be shown in the course of our investigations into the nature of species hybrids that the nature of pure species contradicts this assumption.

Bz, pp. 14–15.

Gaertner: on Hybrids

Ch. xv. "On the origin and formation of hybrid types of plants."

The form and nature of the species is one and the same; hence the former arises from the innermost nature of the plant, and its preservation and propagation depends essentially upon fertilisation. It is modified by foreign fertilising material since pollen possesses a form-determining power as well as an enlivening power. The formation of the types depends upon this formative force and the plants which arise receive this force from the seeds. As hybrid types we understand chiefly the external form, but at the same time we also take into consideration the whole complex of all characters by which a hybrid is distinguished from its parents.

The importance of hybrid form and physiognomy has been brought out by the extending of our studies in hybridisation to a greater number of species in the genus. The formation of hybrid types has now become one of the most interesting and difficult of the subjects which are involved in the study of plant reproduction.

The explanation of how the forms of hybrids originate and are constructed out of the elements and characters of the stem-parents is as important for the plant physiologist as it is for the systematic botanist; whilst for the latter a question of life is also involved. Are there stable species of perfect plants or have they been subjected to change or progressive development (*Fortbildung*) in the course of time, as some naturalists believe? This question has already come under discussion and we have given reasons for speaking in favour of the stability of plant species. Further clarification on this point will be furnished by the study of the origin and formation of hybrid types out of the characters of the stem-parents. (pp. 249–250) . . .

The general similarity of hybrids with their stem-parents can be understood by thinking of the seeds as arising from the mixing which occurs in reproduction and not from pollen alone. However, since very few hybrids show an equal mixing of the characters of both types, but the one factor in the union often preponderates over the other, so the question arises: Which laws govern these modifications in the construction of hybrids? For these types are not vague or the result of chance, on the contrary, they always arise in the same manner and are of the same sorts (*in derselben Art und bei den gleichen Arten sich constant wieder erzeugen*).

Do the laws of hybrid types apply to the individual organs of the plant, or to an individual part, e.g. stems, leaves, etc.? No, they apply to the inner nature of the species. Hence the organs which make up the types of the hybrids must be compared in their totality and their interconnection. For the most part the individuality of the hybrid is expressed in its entire habit, but the flower compared with all other parts of the plant is most frequently and clearly distinguished.

Bz, p. 251.

Gaertner: on F2 Hybrids

Frequently one finds that the state of fertility of the individuals of a hybrid type in the second generation is as variable as it is in those from the original reproduction (F1). We have also remarked that in some fertile hybrids the fertility is increased in the second, third and subsequent generations resulting *from artificial fertilisation by their own pollen*, e.g., in *Dianthus chinensis* × *barbatus*. At the same time the organic constitution and potency of the male organs is gradually restored to a state of perfection as a result of these repeated reproductions, and the resulting hybrids generally approach one or other of the stemparents, the original father or mother (of the cross).

Many more exceptionally fertile hybrid plants propagate themselves with no change of type, like pure species. Hence several botanists are inclined to admit that these hybrids are stable species. As such we have found the following:

Aquilegia atropurpurea × *canadensis*
Dianthus armeria × *deltoides*
Dianthus caesius × *arenarius*

Dianthus superbus × *arenarius*
Dianthus superbus × *caryophyllus*
Dianthus superbus × *pulchellus*
Dianthus chinensis × *barbatus*
Lavatera pseudolbia × *thuriangiaca*
Geum urbanum × *rivale*

but there was always a steady loss of fertility and a general breaking up of the species.

Other fertile hybrids, indeed the majority, give rise to various forms, deviating from the normal type, in the second and subsequent generations, i.e. varieties. These are either unlike the mother hybrid or they deviate from it to a greater or lesser extent, i.e. they degenerate in various ways. Koelreuter and Wiegmann also observed this. . . .

In many fertile hybrids these changes in the second and subsequent generations affect not only the flowers but also the entire habit, even to the exclusion of the flowers, whereby the majority of individuals from one reproduction retain the form of the mother hybrid, a smaller number have become like the stem-mother, and finally an individual here and there has moved closer to the stem-father. We found that *Nicotiana rustica* × *paniculata* and its reciprocal, *Aquilegia vulgaris* × *canadensis* and *Dianthus barbatus* × *chinensis* behaved in this way. But this mode of division of the types is not followed by all fertile hybrids; e.g. in *Lavatera triloba* × *olbia* according to Koelreuter's report several individuals were like the maternal type, others were like the paternal type. . . .

The hybrid union of *Zea mays nana* with the red-seeded variety of *Zea mays major*, which probably ought not to be regarded as different species but as mere varieties, did not like *Pisum* give differently coloured seeds immediately after pollination and fertilisation, but in the second generation. The majority of the seeds were yellow, but there were also reddish, grey and striped seeds. These may perhaps be referred to a mere varietal difference rather than to a specific difference between the two stem-parents.

Bz, pp. 421–424.

Amici: his discovery of Pollen Tubes

Various authors have spoken about the structure of pollen, and various conjectures have been put forward with regard to the internal organisation of the little grains which form this dust, but owing to the small size of these corpuscles which makes dissection impossible, we are still none the wiser. We only know that there is a great variety of external forms and that these often go with the differences between one species and another. But we ignore entirely (the question of) how each pollen grain acts on the stigma in order to introduce the *aura seminalis* which it contains. Geoffroy and Malpighi thought that whole pollen grains, having arrived at the stigma, entered by way of the tubes of the pistil, and were transported to the germ. Bonnet, Duhamel and Gleditsch were not far from this opinion. A few others, like Morland, Hill, etc., imagined that the embryos were to be found in the pollen corpuscles themselves, from whence they came out in order to penetrate and lodge in the ovules. Leaving aside many other hypotheses I will cite just the one which supposes that fertilisation is carried out by means of an irritant action of the *aura seminalis* upon the stigma and transmitted to the germ. However, although I intend to deal with pollen in this article, I have no pretensions about my fitness to discuss the various opinions provided on this subject by these learned men, for I am persuaded that the few observations which are mine will provide but a weak support. My only purpose in proclaiming a singular phenomenon which I have noticed in the pollen of *Portulaca oleracea*, is to excite the curiosity of naturalists who possess good instruments, so that they will pursue this type of research and will provide us with some light on an equally remarkable product.

The extremity of the stigma of *Portulaca oleracea* is covered with very slender transparent hairs, filled with corpuscles of the sap, and I thought it would be interesting to find out if these hairs happened to show any movement in their interior. In fact I was able to satisfy myself that the corpuscles passed from the base of the hairs to the summit, from whence, returning to the base, they began the same circuit again, although rather slowly. Repeating this examination several times, I happened to notice a hair to the tip of which a pollen grain was attached. After a little while this grain suddenly burst, ejecting a sort of gut (*boyau*) which was fairly transparent and after extending to the full length of the hair it appeared to unite laterally with it. Continuing my observation of this new organ which had just appeared, I found that it consisted of a simple tube made of a very delicate membrane, and I was greatly astonished to see that it was filled with little bodies of which one part came out of the pollen grain [into the tube] and the other went back into the grain after having described a circuit of the tube or gut.

Amici, G. B. 1824. "Observations sur diverses espèces de plantes." Ann. Sci. Nat. sér 1, *2*, 65–67. (Trans. of his paper of 1823 in: Memorie di matematica e di Fisica della Societa Italiana della Scienza, residente in Modena, *12*, 234–286.)

Amici: on the Growth of the Pollen Tube

The tubes penetrate into the stigma; of all the facts that one can ascertain for a great number of plants, this is the most certain. But does the prolific liquor diffuse through the interstices of the conducting tissue in order to be carried to the embryo, as M. Brongniart has seen and drawn it? No; the phenomenon is clearly even more curious. It is the tube itself that elongates bit by bit, descends through the style and comes into contact with the nucleus [ovule]. To each ovule there is one tube. Perhaps it will occur to you to ask how, in several plants where the style is very long, the pollen tube can cover such a long distance. The pollen grain is not sufficiently large to house so long a tube. I have also reflected on this problem, and I can only explain the fact of the elongation of the pollen tube, upon which I have not the slightest doubt, by supposing that once it has entered the conducting tissue, the tube receives nourishment and a supply of material from this tissue so that it can extend to the required length.

Amici, G. B. 1830. "Note sur le mode d'action du pollen sur le stigmate; extrait d'une Lettre de M. Amici à M. Mirbel." Ann. Sci. Nat. sér. 1, *21*, 331–332.

Amici: Schleiden's Theory of Fertilisation refuted

Is fertilisation in phanerogamic plants achieved, as Schleiden claims, by means of the extremity of the pollen tube which, penetrating into the integuments of the ovule and pushing back the membrane of the embryo sac, there forms a depression in which it lodges and then produces the genuine embryo?

The special studies which I have carried out on the gourd (*Cucurbita pepo*) have convinced me that in this plant fertilisation takes place in a very different manner. At the assembly of the savants of Padua I showed that the pollen tube penetrates into the neck or tip of the nucellus to a certain depth, but never succeeds in penetrating into the embryo sac which already exists and is visible in the nucellus before the introduction of the pollen tubes into the ovules. Probably the prolific humour, which has been deposited close to or even on the surface of the membrane which forms the embryo sac, is absorbed imperceptibly by this membrane, and thus passes into the interior where it mixes with the fluid of the sac thus completing the act of fertilisaton.

Amici, G. B. 1847. "Sur la fécondation des Orchidées." Ann. Sci. Nat. Botanique, sér. 3, 7, 193. (This paper was read at a congress of Italian scientists at Genoa in 1846.)

Amici: on the Formation of the Embryo in Orchis morio

I have already remarked that the embryo sac contains at its base—a point never reached by the tip of the pollen tube—a granular and white liquid. After fertilisation this liquid condenses and it can be clearly seen that it is enclosed within a new cell which soon subdivides into several cells full of grains. These cells multiply many times thus forming the embryo, which little by little comes to occupy the cavity of the nucellus. At the same time the other part of the embryo sac, that part which has been touched by the pollen tube, elongates above, also by dividing into cells, but into clear cells, arranged one after another in the form of a confervoid filament. This retraces the path taken by the pollen tube, enlarges, passes the opening of the integuments and extends into the interior of the placenta, as I have seen it in *Orchis mascula*. (p. 201)

. . .

Now if you ask me what is the nature of the fertilising action of the pollen tube upon the ovule, I reply without hesitation that I ignore it. It is probable, although one cannot demonstrate it, that the subtle fluid which it contains filters across the membranes into the interior of the embryo sac, and that the mixture of the two fluids of the male and female organs constitutes the material capable of organising itself. It is even possible that the generative faculty resides in the membrane of the embryo sac, and that in order to bring this faculty into activity it must draw in the provenant liquid of the pollen. One can conceive of other interpretations of the phenomenon, but it is not my aim to spend my time speculating and to lose myself in the field of hypotheses. I will add only one fact. It is that in the course of my numerous investigations, I have never found more than one pollen tube within the nucellus, although I have several times encountered two embryo sacs and two embryos fertilised by a single pollen tube.

Amici, G. B. 1847. "Sur la fécondation des Orchidées." Ann. Sci. Nat. Botanique, sér. 3, 7, 202.

APPENDIX TO CHAPTER 3

Darwin: Double Parallel between the Effects of Changed Conditions of Life and of Crossing

Thus we see that when organic beings are placed under new and unnatural conditions, and when hybrids are produced by the unnatural crossing of two species, the reproductive system, independently of the general state of health, is affected by sterility in a very similar manner. In the one case, the conditions of life have been disturbed, though often in so slight a degree as to be inappreciable by us; in the other case, or that of hybrids, the external conditions have remained the same, but the organisation has been disturbed by two different structures and constitutions having been blended into one. For it is scarcely possible that two organisations should be compounded into one, without some disturbance occurring in the development, or periodical action, or mutual relation of the different parts and organs one to another, or to the conditions of life. When hybrids are able to breed *inter se*, they transmit to their offspring from generation to generation the same compounded organisation, and hence we need not be surprised that their sterility, though in some degree variable, rarely diminishes. (p. 227)

It may seem fanciful, but I suspect that a similar parallelism extends to an allied yet very different class of facts.

It is an old and almost universal belief, founded, I think, on a considerable body of evidence that slight changes in the conditions of life are beneficial to all living things. We see this acted on by farmers and gardeners in their frequent exchanges of seed, tubers, etc., from one soil or climate to another, and back again. During the convalescence of animals, we plainly see that great benefit is derived from almost any change in the habits of life. Again, both with plants and animals, there is abundant evidence that a cross between very distinct individuals of the same species, that is between members of different strains or sub-breeds, gives vigour and fertility to the offspring. . . .

Hence it seems that, on the one hand, slight changes in the conditions of life benefit all organic beings, and on the other hand, that slight crosses, . . . give vigour and fertility to the offspring. But we have seen that greater changes, or changes of a particular nature, often render organic beings in some degree sterile; and that greater crosses, . . . produce hybrids which are generally sterile in some degree. I cannot persuade myself that this parallelism is an accident or an illusion. Both series of facts seem to be connected together by some common but unknown bond which is essentially related to the principle of life.

O6e, p. 228.

Naudin: Hypothesis of Segregation

A hybrid plant is an individual in which one finds two different essences united, each having its own mode of vegetation and particular finality; they oppose each other and struggle unceasingly to be free from each other. Are these two essences intimately fused? Do they interpenetrate each other to such a degree that they are contained in equal proportions by every portion of the hybrid plant, . . . It is possible that this is so in the embryo and perhaps in the first phases of the development of the hybrid; but it seems much more likely to me that the latter, at least in the adult state, is an aggregation of heterogeneous portions which, when considered separately, are homogeneous and unispecific. But these portions are distributed equally or unequally between the two species, and are intermingled in various proportions in the plant organs. The hybrid, according to this hypothesis, would be a living mosaic, the discordant elements of which the eye cannot perceive as long as they remain intermingled; but if, on account of their affinity, the elements of the same species approach one another and are aggregated into quite considerable masses, it could result that parts are formed discernible to the eye, sometimes whole organs, as we see in *Cytisus Adami*, the oranges and citron hybrids of the Bizzaria group, *Datura Stramonio-laevis*, etc. It is this more or less visible tendency of the two specific essences to free themselves from their union which has led several hybridists to say that hybrids resemble the mother in their foliage, their father in the flowers, or *vice versa*. It did not escape that ingenious experimenter Sageret, who found hybrids less remarkable for the state of intermediacy of each of their organs than for the pronounced resemblance of certain organs with those of the father, and of those of certain others with the mother. He even cites a hybrid of the cabbage and the radish, of which certain siliquas [fruits] were those of the cabbage and others those of the radish. If he has not mistaken a monstrosity for a hybrid he has added a remarkable example of disunited hybridity (*hybridité disjointe*) to those which we know.

Although the facts may still be insufficient to draw a definite conclusion, it seems that the tendency of species to separate themselves, or if you like, to localise themselves on different parts of the hybrid increases with the age of the plant, and that it is more and more pronounced as the vegetable body approaches its termination, which is on the one side the production of the pollen, and on the other the formation of the seed. Effectively it is at the organic summit of the hybrids, in the neighbourhood of the organs of reproduction, that these segregations (*disjonctions*) become more manifest: in *Cytisus Adami* segregation takes place on the flowering branches. It occurs on the fruits themselves in the Bizzaria orange and in *Datura Stramonio-laevis*. In *Mirabilis longifloro-Jalapa* and *Linaria purpurea* it is the corolla which shows the phenomenon of segregation by the separation of the pure (*propre*) colours into the species concerned. These facts permit one to think that the pollen and ovules, the pollen above all which is the extreme term of the male flowering, are precisely the parts of the plant where specific segregation takes place with the most energy. And what makes this hypothesis more probable is that these organs are at the same time very elaborate and very small, a double reason for bringing about a more perfect localisation of the two essences. If we admit this hypothesis, and I acknowledge that it seems extremely probable, then all the changes which happen in the hybrids of the second and later generations explain themselves. On the other hand they would be inexplicable were one to deny it.

Let us suppose, in the *Linaria* hybrid of the first generation, that segregation takes place at the same time in the anther and in the contents of the ovary; that some of the pollen grains belong totally to the paternal species, some entirely to the maternal species; that in others segregation has not occurred or is only beginning. Let us admit further that the ovules are segregated to the same degree in the sense of the father and of the mother. What will happen when the pollen tubes descend into the ovary and seek out the ovules in order to fertilise them? If the tube of a pollen grain reverted (*revenu*) to the paternal species encounters an ovule segregated in the same sense, a *perfectly legitimate* fertilisation will occur, the result of which will be a plant reverted entirely to the paternal species. The same combination taking place between a pollen grain and an ovule both segregated in the sense of the mother of the hybrid, the product will revert by itself into the species of the latter. On the other hand, if the combination takes place between a pollen grain and an ovule both segregated in opposite senses, a true *cross fertilisation* is brought about, like that which gave birth to the hybrid itself, and the result will be a form still intermediate between the two specific types. The fertilisation of an ovule not segregated, by a pollen grain segregated in one sense or the other would give a quarteron [i.e. a back-crossed hybrid, see Table II, p. 68]; and since segregations can take place in all degrees as much in the pollen as in the ovules, combinations will arise which chance alone directs. These would give rise to that multitude of forms which we have seen produced in the *Linaria* hybrids and in *Petunia*, from the second generation onwards.

Naudin, C. 1863. "Nouvelles recherches sur l'hybridité dans les végétaux." Ann. Sci. Nat. Botanique, sér. 4, *19*, 191–194.

Galton: Stability of Type

I will now explain what I presume ought to be understood, when we speak of the stability of types, and what is the nature of the changes through which one type yields to another. Stability is a word taken from the language of mechanics; it is felt to be an apt word; let us see what the conception of types would be, when applied to mechanical conditions. It is shown by Mr. Darwin, in his great theory of *The Origin of Species*, that all forms of organic life are in some sense convertible into one another, for all have, according to his views, sprung from common ancestry, and therefore A and B have both descended from C, the lines of descent might be remounted from A to C, and redescended from C to B. Yet the changes are not by insensible gradations; there are many, but not an infinite number of intermediate links; how is the law of continuity to be satisfied by a series of changes by jerks? The mechanical conception would be that of a rough stone, having, in consequence of its roughness, a vast number of natural facets, on any one of which it might rest in "stable" equilibrium. That is to say, when pushed it would somewhat yield, when pushed much harder it would again yield, but in a less degree; in either case, on the pressure being withdrawn it would fall back into its first position. But, if by a powerful effort the stone is compelled to overpass the limits of the facet on which it has hitherto found rest, it will tumble over into a new position of stability, whence just the same proceedings must be gone through as before, before it can be dislodged and rolled another step onwards. The various positions of stable equilibrium may be looked upon as so many typical attitudes of the stone, the type being more durable as the limits of its stability are wider. We also see clearly that there is no violation of the law of continuity in the movements of the stone, though it can only repose in certain widely separated positions.

Galton, F. 1869. *Hereditary Genius*. (Reprinted in: H.G., pp. 421–422.)

APPENDIX TO CHAPTER 4

Mendel: the Variability of Cultivated Plants

. . . The opinion has often been expressed that the stability of the species is greatly disturbed or entirely upset by cultivation, and consequently there is an inclination to regard the development of cultivated forms as a matter of chance devoid of rules; the colouring of ornamental plants is indeed usually cited as an example of great instability. It is, however, not clear why the simple transference into garden soil should result in such a thorough and persistent revolution in the plant organism. No one will seriously maintain that in the open country the development of plants is ruled by other laws than in the garden bed. Here, as there, changes of type must take place if the conditions of life be altered, and the species possesses the capacity of fitting itself to its new environment. It is willingly granted that by cultivation the origination of new varieties is favoured, and that by man's labour many varieties are acquired which, under natural conditions, would be lost; but nothing justifies the assumption that the tendency to the formation of varieties is so extraordinarily increased that the species speedily lose all stability, and their offspring diverge into an endless series of extremely variable forms. Were the change in the conditions the sole cause of variability we might expect that those cultivated plants which are grown for centuries under almost identical conditions would again attain constancy. That, as is well known, is not the case, since it is precisely under such circumstances that not only the most varied but also the most variable forms are found. It is only the *Leguminosae*, like *Pisum*, *Phaseolus*, *Lens*, whose organs of fertilisation are protected by the keel, which constitute a noteworthy exception. Even here there have arisen numerous varieties during a cultural period of more than 1,000 years under most various conditions; these maintain, however, under unchanging environments a stability as great as that of species growing wild.

It is more than probable that as regards the variability of cultivated plants there exists a factor which so far has received little attention. Various experiments force us to the conclusion that our cultivated plants, with few exceptions, are *members of various hybrid series*, whose further development in conformity with law is varied and interrupted by frequent crossings *inter se*. The circumstance must not be overlooked that cultivated plants are mostly grown in great numbers and close together, affording the most favourable conditions for reciprocal fertilisation between the varieties present and the species itself.

Mendel, G. 1956. *Experiments in plant hybridisation*. Harvard University Press, Cambridge, Mass. pp. 31–32.

Darwin: Hypothesis of Pangenesis

I have now enumerated the chief leading points which we naturally wish to connect together by some intelligible bond. It will, I presume, be admitted that the protoplasm or formative matter, included within the germ and male element, and endowed with vital force, causes in seminal generation the development /* of each new being whose germs and buds agree, as we have seen, in structure as far as this is visible, in many remarkable attributes, as in varying, inheritance, reversion, and hybridisation, and lastly in their fully developed product. Hence it seems by far the simplest belief that protoplasm, identical in nature with that within the germ, collects at certain points to form buds. If this view be admitted it must certainly be extended to fissiparous generation, to the renewal of an amputated limb, to the healing of a wound and probably to continuous growth. We are thus led to believe that protoplasm of the same nature, must be diffused throughout the whole of each organic being, ready when superabundant to form by budding new beings, both at the period of maturity and in the cases of alternate generation during youth; and ready to form new structures as after inflammation, and ready to repair lost or wasted structures. On this view we must believe that the reproductive organs do not by any means exclusively form the generative protoplasm, if indeed they form any of it, but only select and accumulate it in the proper quantity, and make it ready for separate existence. /

We can thus understand the antagonism that has long

* The oblique strokes show the end of each page of the manuscript.

been observed in plants between increase by buds, rhizomes, suckers and seminal generation (and indeed between the latter and active growth during youth); for in both cases the same protoplasmic matter is consumed; and there is not enough for both methods of propagation. It is surprising that this antagonism should be as general as it is; but does not invariably hold good; for the young males of the salmon, whilst very small, have their reproductive organs active; and Ernst Haeckel has recently (*Monatsbericht Akad. Wiss. Berlin.* Feb. 2, 1865) described the wonderful case of a medusa, with its reproductive organs active, which at the same time produces by budding a widely different form of medusa, which likewise has the power of seminal reproduction.

Furthermore, I am led to believe from analogies immediately to be given that the protoplasm or formative matter which is diffused throughout the whole organisation, is generated by each different tissue and cell or aggregate of similar cells;—that as each tissue or cell becomes developed, a superabundant atom or gemmule as it may be called of the formative matter is thrown off;—that these almost infinitely numerous and infinitely minute gemmules unite together in due proportion / to form the true germ;—that they have the power of self-increase or propagation; and that they here run through the same course of development, as that which the true germ, of which they are to constitute elements, has to run through, before they can be developed into their parent tissue or cells. This may be called the hypothesis of Pangenesis.

On this hypothesis the many different parts of the structures of each individual may be compared to so many distinct organic beings, united together, but each of which propagates its own proper form. The union is far more intimate than that of flower buds or leaf-buds on the same tree, or of the polypi on the same coral; / but even in these cases we have some differentiation in the so-called individuals, and some parts in common; for plants have trunks and roots in common, and some kinds habitually produce two kinds of flowers; and the polypi of some corals have certain parts and the power of movement in common. /

Olby, R. C. 1963. "Charles Darwin's manuscript of Pangenesis." British Journal for the History of Science, *1*, pt. 3, 258–259. Reprinted by kind permission of the acting Editor.

Dzierzon: Hybridisation of Bees yielding a 1 : 1 ratio for the Drones

. . . If she [queen bee] herself originates from a hybrid brood, it is impossible for her to produce pure drones, but she produces half Italian and half German drones, but strangely enough, not according to the type but according to number, as if it were difficult for nature to fuse both species into a middle race.

Dzierzon. 1854. "Die Drohnen." Der Bienenfreund aus Schlesien. Herausgegeben von Pfarrer Dzierzon in Carlsmarkt. Brieg. August, No. 8, 63–64. Translation taken from: Zirkle, C. 1951. "Gregor Mendel and his precursors." Isis, *42*, 102.

APPENDIX TO CHAPTER 5

*Gregor Mendel's Autobiography**

Praiseworthy Imperial and Royal Examination Commission!

In accordance with the high regulations of the Ministry of Public Worship and Education, the respectfully undersigned submits a short sketch of his life.

* Translated from the German by Mrs. Hugo Iltis, and reprinted by kind permission of the Managing Editor of the *Journal of Heredity*.

The same was (in accordance with enclosure A) born in the year 1822 in Heinzendorf in Silesia, where his father was the owner of a small farm. After he had received elementary instruction at the local village school, and later at the Piarist's College [upper elementary school] in Leipnik, he was admitted in the year 1834 to the first grammatical class of the Imperial Royal Gymnasium in Troppau. Four years later, due to several successive disasters, his parents were completely unable to meet the expenses necessary to continue his studies, and it therefore happened that the respectfully undersigned, then only sixteen years old, was in the sad position of having to provide for himself entirely. For this reason, he attended the course for "School Candidates [applicants] and Private Teachers" at the district Teacher's Seminary in Troppau. Since, following his examination, he was highly recommended in the qualification report (enclosure B), he succeeded by private tutoring during the time of his humanities studies in earning a scanty livelihood.

When he graduated from the Gymnasium in the year 1840, his first care was to secure for himself the necessary means for the continuation of his studies. Because of this, he made repeated attempts in Olmütz, to offer his services as a private teacher, but all his efforts remained unsuccessful because of lack of friends and recommendations. The sorrow over these disappointed hopes and the anxious, sad outlook which the future offered him, affected him so powerfully at that time, that he fell sick and was compelled to spend a year with his parents to recover.

In the following year the respectfully undersigned found himself finally placed in the desired position of being able to satisfy at least his most necessary wants by private teaching in Olmütz, and thus to continue his studies. By a mighty effort, he succeeded in completing the two years of philosophy (enclosures D, E, F, G). The respectfully undersigned realised that it was impossible for him to endure such exertions any further. Therefore, after having finished his philosophical studies, he felt himself compelled to step into a station of life, which would free him from the bitter struggle for existence. His circumstances decided his vocational choice. He requested and received in the year 1843 admission to the Augustinian Monastery St. Thomas in Altbrünn.

Through this step, his material circumstances changed completely. With the comfortableness of his physical existence, so beneficial to any kind of study, the respectfully undersigned regained his courage and strength and he studied the classical subjects prescribed for the year of probation with much liking and devotion. In the spare hours, he occupied himself with the small botanical-mineralogical collection which was placed at his disposal in the monastery. His special liking for the field of natural science deepened the more he had the opportunity to become familiar with it. Despite his lack of any oral guidance in these studies, plus the fact that the autodidactic method here, as perhaps in no other science, is extremely difficult and leads to the goal only slowly, he became so attached to the study of nature from this time on that he will not spare any effort to fill the gaps that are still present through self instruction and the advice of experienced men. In the year 1846, he also attended courses in agriculture, pomiculture, and wine-growing at the Philosophical Academy in Brünn (enclosure, H, I, K).

After completing the theological studies in 1848, the respectfully undersigned received permission from his prelate to prepare himself for the philosophical rigorosum [examination for the Doctor of Philosophy degree]. In the following year at the time when he was about to undergo his examination, he was asked to accept the position of a substitute teacher at the Imperial Royal Gymnasium in Znaim, and he followed this call with pleasure. From the beginning of his substitute teaching, he made all efforts to present his assigned subjects to the students in an easily comprehensible manner. He hopes his endeavour was not quite without success since, during that private tutoring to which he owed his bread for four years, he found sufficient opportunity to collect experiences regarding the possible accomplishments of the students and the different grades of their mental capacity.

The respectfully undersigned believes to have rendered with this a short summary of his life's history. His sorrowful youth taught him early the serious aspects of life, and taught him also to work. Even while he enjoyed the fruits of a secure economic position, the wish remained alive within him to be permitted to earn his living. The respectfully undersigned would consider himself happy if he could conform with the expectations of the praiseworthy Board of Examiners and gain the fulfilment of his wish. He would certainly then shun no effort and sacrifice to comply with his duties most punctually.

Znaim, on the 17th April 1850

Gregor Mendel

Subst. Professor on the Imp. Roy. Gym. in Znaim

Iltis, Mrs. Hugo. 1954. "Gregor Mendel's Autobiography." J. Hered. *45*, 231–234.

Unger: on the Origin of Species

In the following passage on the origin of species Unger again expresses ideas which are clearly derived from the Nature-philosophers' teaching. Thus he speaks of the whole plant kingdom as an "edifice", i.e. a unity. Nevertheless, his views on the origin of variation and on the instability of species are well worth reading. Note that he speaks of hybrids as "entirely new species as it were arising from the combination of two pre-existing species", but that he reckons them insufficient to form species on account of their lack of duration "so that such bastards are never in a condition to dispute and attain their citizenship among the other species of plants born their equals".

If we conceive the species, as has been customary hitherto, as an aggregate of similarly-formed (similarly-natured) individuals—in which not a single quality permanently alters (immutable characteristics), as the experience of our observation shows—we are inevitably impelled, in the explanation of this question, to the conclusion that the origin of a species could not possibly have taken place from any of its precursors. There is nothing for it, therefore, but to assume that forces beyond the pale of the organic world co-operate in the production of the species—an assumption which, if not in actual contradiction to the universal operation of inorganic forces, yet sounds at least like a miracle.

Far otherwise does the matter present itself, if, following the track of analogy, we regard the species as a sum-total of elements capable of production, and therefore of alteration; in which indeed no metamorphoses are to be perceived, except in lengthened periods, but in which, within the compass of many centuries (wherein it may be certainly computed that the generations of existence of every organic being can be comprised), the germination, growth, blossoming, fructification, and ripening, of the species follow none the less.

It would, however, be erroneous to assume that the diversity of species consisted only in this process of metamorphosis; but who can deny that new combinations of the elements arise out of this permutation of vegetation ever reducible to a certain law—combinations which emancipate themselves from the preceding characteristics of the species, and appear as new species? I must not be asked "When?"—nor how such offshoots from the already-existing species arose. On these points, nothing but the history of the development of the whole plant-world can possibly afford a solution. But this much is clear—that this change of generation relating to species can belong neither to the youth nor to the old age of the species, but to the period of its greatest strength, its highest development, as well in extent as in energy of vegetation.

Nevertheless, phenomena strike us, even in our fragmentary term of observation, which are significant in supporting the above views, and which, even if they do not, as was supposed, invalidate the theory of stability of species, still clearly reveal the great process of metamorphosis of one species into another, and consequently the comprehension of these within a higher unity. These phenomena are such as belong partly to normal life, partly to morbid and uncontrolled vegetation. The deviation of particular characteristics from the normal condition in the succession of generations is one of the commonest phenomena. According to the greater or less permanence of these deviations, we call the one a variation (*variatio*), the other formation of race. To what an extent these often proceed our garden plants show, in which we are scarcely able, often quite unable, to recognise the progenitors. That these deviations arise not altogether from an alteration of outward influences, such as from a change of light, air, moisture, soil, or so on, is demonstrated by the fact, that two similar kinds of plants frequently become altogether different under these circumstances.

Whilst the vegetation of both is equally strongly affected, it is arrested in one, whilst it produces no effects on the other. The endeavour, therefore, to trace the diversities of species to the effect of outward influences, such as the nature of the soils, assuredly misses the true cause. Equally insufficient, though not without significance, proves the effect which the reproductive activity of one kind of plants exerts over the other, whereby in the higher, as well as in the lower growing plants, even in mosses and ferns, arise hybrids, entirely new species as it were arising from the combination of two pre-existing species. Their duration, although lasting some generations, is, nevertheless, always short, so that such bastards are never in a condition to dispute and attain their citizenship among the other species of plants born their equals.

Finally there remain in the balance the phenomena of abnormal vegetation, as not unimportant influences in the constant presence of a transformative plant-growth. Who is not familiar with the signs of transformed vegetation which meet him in every meadow, in every garden? Not only do stem and leaves expand to an excessive degree, a different texture, other constituents, etc., appear; even into the one so regular order of the leaves diversity enters, the cycles alter, the succession of formations is disturbed, and transformations of the strangest kind present themselves. To whom are the so-called double flowers, perfoliate blossoms, incised fruits, and so on, unknown? It is, in all cases, the impatient vegetation which here concealed, there openly, produces these phenomena. . . .

And how could this spirit of change, this representative of the unconstant, of the transitory, fail to transgress the narrow bounds of peculiarity of species? It were scarcely credible. If then we must dismiss as incorrect all previous observations on the changes of types of species, we yet cannot avoid recognising, in the genius which marks the species, seeks to preserve its unity through all times and localities, and does in truth preserve it, the strength which not only converts water into wine, but is able, with similar magic power, to transmute also one species into another. But if all distinctions of species sink into nothing before this magic wand, how can it be doubted, that in the higher categories, the same generic unity reigns, that they likewise are but the result of propagation in distant zones? We should much err if we did not ascribe a *real* existence to these unities included in one general view by the mind. If the universal unity of the plant-body is rendered possible only by the production of all its single elements one out of the other, then is this unity in the whole creation of the plant-world, assuredly in like manner possible only by the originating of one member from another, one species from another, one genus, one family from another. And as in the plant-body, not even a single cell can be produced from any extrinsic source, equally impossible is it for a species, a genus, an order, etc., of plants, to be produced from any extrinsic source, and not to have proceeded from a previous one.

Thus rises up to our astonished gaze not only the wonderfully-proportioned structure of the visible plant-form, but this, itself, extends into regions to which our mortal eye is no longer able to penetrate. Not only the individual plant, but the whole plant-kingdom is an edifice—an edifice for which the thousands and thousands of parts, as leaves and flowers and single cells, serve as building-stones.

Unger, F. J. A. N. *Botanical letters to a friend translated by B. Paul.* London. 1853, pp. 93–95.

Fisher: Statistical Analysis of Mendel's Results

The scores for round and wrinkled F2 seeds in Mendel's first cross were: round 5,474 : wrinkled 1,850. Fisher's comment is: "The deviation from the expected 3 : 1 is less than its standard error of random sampling." For the scores, yellow seeds 6,022 : green 2,001, he says: "The agreement with expectation is here even closer."* He goes on to calculate the χ^2 values for the 3 : 1 ratios found by Mendel in the case of the 7 character differences, and he finds that the probability of exceeding the observed deviations is ·95;† i.e., if the experiments were repeated 100 times the agreement with the 3 : 1 ratio in 95 of them would be less than that obtained by Mendel.

If we repeat the χ^2 test for the character pair yellow-green seeds we arrive at a value for p slightly less than Fisher's ·95 for the 7 pairs of contrasted characters. The

* Fisher, R. A. 1936. "Has Mendel's work been rediscovered?" Ann. Sci., *1*, 121.

† Ibid., Table V, p. 131.

same test applied to the results of other hybridists gives the following values for p:

Hybridist	*Year*	*Yellow* *Obs.*	*Yellow* *Exp.*	*Green* *Obs.*	*Green* *Exp.*	P.
Mendel	1865	6,022	6,017	2,001	2,006	0·9
Tschermak	1900	3,580	3,577	1,190	1,193	0·9
Hurst	1904	1,310	1,316	445	439	0·7–0·8
Correns	1900	1,394	1,386	453	462	0·5–0·7
Darbishire	1909	109,060	108,935	36,186	36,311	0·5

Now why are Tschermak's results also too good to be true? The answer, it seems to me, is a very simple one. Both Tschermak and Mendel stopped scoring their results when the totals gave a striking confirmation of a simple ratio. They did not alter their results, nor did they classify "difficult" seeds with whichever group needed to be enlarged. They merely allowed the idea of a simple ratio between the scores to influence the point at which they stopped scoring. And this is not a question of obtaining very good results at first followed by bad results which one excludes.

Let us assume, with Fisher, that in Mendel's experiments with seed characters the hybrid plants were harvested whole and hung up to dry. In the winter months their pods were shelled. The totals for each plant were recorded one by one. As more and more plants were dealt with, the grand totals grew. When 600 seeds had been harvested we may presume that segregation was of the order of, say, 140 green to 460 yellow (theoretical expectancy, 150 : 450). If Mendel had not already thought out his factorial theory he must surely have perceived the simple relationship between the totals of the segregating classes at this point. He would then have gone on to achieve as convincing a demonstration of this ratio as his material would allow.

Now, if one were to note the totals of the segregating

classes each time the scores of individual plants were added, then one would see how they fluctuated this way and that from the 3 : 1 ratio. It is a simple matter to look out for a point in the series of mounting totals where deviations from expectancy in either direction tend to cancel each other out. If one elects to stop at such a point the result will be better than one would expect on the basis of the χ^2 test.

If Mendel stopped recording his seeds before he had exhausted the material one would expect that his totals would be less than that of an average crop for the population of mother plants grown. This is so. Mendel stated that fully ripe pods contained between 6 and 9 seeds. If we take 6 as the average number, in order to make an allowance for unripe pods, then the 7,324 seeds which Mendel harvested from 253 plants would have come from 1,046 pods, thus giving 4 to 5 pods per plant.

When we turn to the numbers Mendel recorded for the colour of the unripe pod we find a very different state of affairs. He grew only 580 plants, and since he was dealing with a character of the mother plant his total score could not exceed 580. The segregation he recorded of 428 : 152 gives a χ^2 value of 0·449 and probability of 0·5. His approximation to the 3 : 1 ratio is closer than this in the case of position of flowers (858 records), length of stem (1,064 records), and form of pod (1,181 records), but not as close as the results for seed characters. These facts also support the contention that it was only when Mendel could choose where to stop scoring that his results appear to be too good to be true; i.e. where he had a very large population at his disposal.

When we bear in mind the fact that Mendel carried out this work over 100 years ago, it seems unlikely that he was greatly concerned about the statistical significance of his results, or about the bias which he may have shown in favour of concluding his scoring at an advantageous point. Of course he knew about the sampling error, for it was because all his precursors had raised small populations of hybrids that they failed to note the 3 : 1 ratio. Mendel illustrated this point by giving the scores for the round and wrinkled and green and yellow seeds of 20 individual plants. This done, he gave no further individual totals. He saw no need to do so, for he had made his point and he was not concerned with the probable error of his results. His concern was to obtain a striking demonstration of the ratio he had discovered earlier in the process of counting.

Mendel: Geum urbanum × G. rivale

The notes which Mendel made in the flyleaf at the back of his copy of Gaertner's book, *Versuche und Beobachtungen über die Bastarderzeugung im Pflanzenreich*, are produced in Plate 10. The six pairs of contrasted characters which he listed translate as follows:

G. urbanum	*G. rivale*
a. lower joint of awn glabrous.	hairy at base.
b. lower joint of awn four times as long as upper joint.	almost as long as upper joint.
c. upper joint pubescent at the base.	pubescent to just under the tip.
d. flowers erect.	flowers nodding.
e. fruit-bearing calyx reflexed.	erect.
f. carporphore lacking.	carporphore almost as long as the calyx.

The letters above this list: $\frac{\text{ABcDEe}}{\text{ABcdEe}}$ do not appear to represent the genotype of one hybrid, but of two which differ from each other in the character difference D d, but agree in being homozygous for A, B, and c, and heterozygous for E. With the exception of the character difference, erect to nodding flowers, the hereditary transmission of none of these traits has been followed. It would be interesting to find out if any of them show clear 3 : 1 or 1 : 2 : 1 segregation in the F2 generation.

Mendel: Phaseolus multiflorus × P. nanus

A page from Mendel's notebook which is thought to refer to colour inheritance in *Phaseolus* is reproduced in Plate 11. The figures in the left-hand column are clearly the theoretical frequencies for a di-hybrid cross since they can be reduced to 1 : 1 : 2 : 4 : 4 : 4, which adds up to 16 just as does 9 : 3 : 3 : 1. The symbols indicate the colour types rather than genetic factors, for he introduces three different symbols and no corresponding allelomorphs, instead of simply two pairs of allelomorphs. In the second column he regroups these colour types and again gives expected frequencies. His experimental results are recorded in the third column, and beside them the deviation from the expected values. Finding such a large discrepancy between observed and expected frequencies he tries other groupings of the various colours encountered. Thus:

343 light violet and violet.
92 blue.
166 white.

These are the figures which Fisher suggested may represent a 9 : 3 : 4 segregation.* But Mendel tried to fit them to a 7 : 2 : 3 ratio. Clearly he failed to find a satisfactory explanation. Nor can we unreservedly accept Fisher's explanation, since one would not expect to find yellow offspring in the cross of the scarlet runner and dwarf bean. The colours listed suggest Mendel's *Linaria vulgaris* × *L. striata* cross more strongly than they do his *Phaseolus* cross. Hence the chief interest of this manuscript lies in the demonstration it affords of Mendel trying to fit observation and theory.

* Darlington, C. D. & Mather, K. 1952. *The elements of genetics*. London. p. 8.

Mendel: Notes on Pisum *in his copy of Gaertner's book*

These notes which are reproduced in Plate 12 translate as follows:

Pisum arvense: flowers solitary, wings red.

Pisum arvense et sativum: pods almost cylindrical, in *Pisum umbellatum* Mill. cylindrical and straight; in *saccharatum* Host. straight, ensiform, constricted on both sides. (var. *flexuosum* Willd. sickle-shaped, seeds small, angular); in *Pisum quadratum* Mill. straight, ensiform, not constricted, seeds pressed tightly together. In *Pisum sativum* and *arvense*, the bases of the stipules rounded and denticulate-crenate, stipules cordate. In *saccharatum* and *quadratum*, stipules obliquely incised, pods pressed flat. In *sativum*, *saccharatum* and *umbellatum*, seeds round.

These notes are important because they show Mendel at work, hunting for clearly-marked character differences between the various forms of peas. Hence it is reasonable to assume that these notes were written prior to the purchase of the 34 varieties of peas for testing in 1854. If this be so, Mendel must have purchased his copy of Gaertner's book, *Versuche und Beobachtungen über die Bastarderzeugung im Pflanzenreich*, before he had worked out the detailed plan of his experiments with *Pisum*.

Clearly no definite conclusion can be drawn on this point, but I hold it as very probable that Mendel learnt of Gaertner's work from Unger in Vienna, that Mendel looked at the book during his botanical studies there in 1852, and that he subsequently purchased a copy which he read in more detail in 1853–4, before he chose the 34 varieties of peas. At the back of the book he noted down some of the character differences between the various varieties and species of *Pisum*, presumably from published descriptions. Later he chose two of these distinguishing characters for the experiments reported in 1865; i.e. form of the seed angular or round (character difference No. 1 of the 1865 paper), and shape of the pod constricted or inflated (character difference No. 4 of the 1865 paper).

APPENDIX TO CHAPTER 6

Mendel: the Process of Segregation

. . . If the reproductive cells be of the same kind and agree with the foundation cell of the mother plant, then the development of the new individual will follow the same law which rules the mother plant. If it chance that an egg cell unites with a *dissimilar* pollen cell, we must then assume that between those elements of both cells, which determine opposite characters, some sort of compromise is effected. The resulting compound cell becomes the foundation of the hybrid organism, the development of which necessarily follows a different scheme from that obtaining in each of the two original species. If the compromise be taken to be a complete one, in the sense, namely, that the hybrid embryo is formed from two similar cells, in which the differences are *entirely and permanently accommodated* together, the further result follows that the hybrids, like any other stable plant species, reproduce themselves truly in their offspring. . . .

With regard to those hybrids whose progeny is *variable* we may perhaps assume that between the differentiating elements of the egg and pollen cells there also occurs a compromise, in so far that, the formation of a cell as foundation of the hybrid becomes possible; but, nevertheless, the arrangement between the conflicting elements is only temporary and does not endure throughout the life of the hybrid plant. Since, in the habit of the plant, no changes are perceptible during the whole period of vegetation, we must further assume that it is only possible for the differentiating elements to liberate themselves from the enforced union when the fertilising cells are developed. In the formation of these cells all existing elements participate, in an entirely free and equal arrangement, by which it is only the differentiating ones which mutually separate themselves. In this way the production would be rendered possible of as many sorts of egg and pollen cells as there are combinations possible of the formative elements.

Mendel, G. 1956. *Experiments in plant hybridisation.* Harvard University Press. Cambridge, Mass. pp. 36–37.

(It is worth comparing this passage with the very similar passage from Naudin. See Appendix to Chapter 3, pp. 167–170.)

Tschermak: on de Vries' Attitude to Mendel

Speaking of de Vries in 1895, at the time of his discovery of the segregating proportions of the F2 generation of his cross: *Oenothera lamarckiana* × *brevistylis*, Tschermak said:

At this time he reads Professor Bailey's book* *Plant Hybridisation* which the author sent him in 1892, and he finds there the Mendel citation from Focke. He procures a copy of Mendel's paper and is greatly astonished to find the same regularities described in detail and explained by Mendel. . . .

It is interesting that de Vries evidently became somewhat jealous of the rapid development of Mendelism and he considered his Mutation Theory rather neglected, especially by breeders. Only thus can one account for his failure to mention Mendel's name even once in his book *Pflanzenzüchtung* in 1907, and his brusque refusal when invited to sign the petition for the erection of a Mendel memorial in Brünn in 1908.

Tschermak-Seysenegg, E. von. 1951. *Historischer Rückblick auf die Wiederentdeckung der Gregor Mendelschen Arbeit.* Verh. zool.-bot. Ges. Wien, 92, 30–31.

* Actually it was Bailey's paper, not his book. See p. 128.

(This is a fresh translation but the whole paper has been translated before in: J. Hered., *42* (1951), 163–171.)

Tschermak: the Achievement of Rediscovering Mendel's Laws

The three rediscoverers were perfectly in agreement that the independent discovery of the laws of inheritance in 1900 was far from being the achievement that it was in Mendel's day; for the studies which had appeared meanwhile, especially the cytological investigations of Hertwig, Strasburger and others, had made the task much easier. It meant less to them to be celebrated as the rediscoverers of regularities which they themselves termed Mendel's laws than for their employment of the Mendelian theory for the development of their own quite different fields of research, de Vries for the Mutation Theory, Correns for basic research into heredity, particularly the heredity of sex, and I for practical plant breeding.

Tschermak-Seysenegg. 1951. *Opus cit.*, p. 34.

Tschermak's Part in the Rediscovery

When some of Correns' and Fritz Wettstein's students tried to establish a gradation in merit between de Vries, Correns and Tschermak with respect to their understanding of the significance of Mendel's work, Tschermak objected. He had, he declared, at once recognised the importance of Mendel's laws, for in 1900:

. . . I had applied for the acceptance and reprinting of Mendel's paper in *Ostwald's Klassiker der exakten Wissenschaften*, but owing to the hesitation expressed by Solms-Laubach, it was not until 1901, after yet another energetic request on my part, that it was published. Meanwhile, Mendel's paper was reprinted in *Flora* in 1900 (!) at the suggestion of Goebel. It was not easy for the young Tschermak to establish his part in the discovery of Mendelism and in its utilization for practical breeding, for only the names of de Vries and Correns found a place in the then leading textbooks. This omission, however, was corrected in the later editions.

Tschermak-Seysenegg, E. von. 1937. "Erinnerungen an die Wiederentdeckung der Mendel'schen Vererbungsgesetze vor 37 Jahren". Züchter, *9*, 145–146.

Dates of the Introduction of some Terms used in Genetics

Allelomorph	Bateson	1902
Chromosome	Waldeyer	1888
Dominant	Mendel	1866
F_1, F_2, F_3 . . .*	Bateson & Saunders	1902
Factor†	Gaertner	1849
Factor‡	Mendel	1866
	de Vries	1889
	Bateson	1901
Gamete	Strasburger	1877
	Bateson	1902
Gametic Coupling	Bateson, Saunders & Punnett	1906
Gene	Johannsen	1909
Genetics	Bateson	1906
Genotype	Johannsen	1909
Heterozygote	Bateson	1902
Homozygote	Bateson	1902
Hybrid§	Mendel	1866
Linkage	Morgan	1910
Pangene	de Vries	1889
Recessive	Mendel	1866
Zygote	Strasburger	1877
Zygospore	de Bary	1858

* The idea of this notation was suggested to Bateson by the system of letters which Galton used to represent family relationships in his book *Hereditary Genius*, 1869.

† Gaertner used this word to signify the whole contribution of one parent to the heritage of the offspring.

‡ Mendel used this word in the sense of the determinant, but on one occasion only. Elsewhere he either used the word elements or simply referred to the characters as if they were the determinants. (For an excellent account of this point see: Darlington. 1964. *Genetics and Man.* p. 94.)

§ The word hybrid (ὕβρις = outrage, rape) came to be used chiefly for inter-specific crosses, but Mendel's use of it for his cross-bred peas, together with Darwin's denial of any fundamental distinction between species hybrids and varietal cross-breds, ensured it a wider usage in this country. In Germany, however, the term most widely used is still "Bastard".

Postscript

Through the kindness of the Publisher the following résumé of further information found since the writing of this book is included here.

On Gaertner and Mendel

Sir Gavin de Beer gave an excellent account of the attitude of the authorities of the Austro-Hungarian Empire to the evolution hypothesis in the years which followed the 1848 revolution, in his paper, "Mendel, Darwin, and Fisher" (*Notes & Records of the Roy. Soc.*, *19* (1964) 192–231). This part of the paper is largely based on a little-known paper by B. Matoušková, "The beginnings of Darwinism in Bohemia", which appeared in *Folia Biologica, Praha*, *5* (1959) 169–182.

Sir Gavin also drew attention to passages in Mendel's "Versuche über Pflanzenhybriden", which he claims represent a direct cirticism of Darwin's views on variation. Mendel must have known about Darwin's work in 1865 for it was expounded with much enthusiasm by Mendel's friend, Alexander Makowsky, at the meeting of the Brünn Scientific Society which preceded the February 8 reading of part one of Mendel's "Versuche". Nevertheless, it seems more probable that the chief passage in question (reprinted here in the Appendix, pp. 171–172) was aimed primarily at Gaertner and Koelreuter. They held that variability is in the main confined to cultivated plants and is therefore in some way due to the unnatural conditions and treatment they receive. Hence these authors felt justified in upholding the doctrine of the fixity of species, despite the existence of some contrary results from their experiments. On the subject of transmutation Gaertner faithfully recorded the opinions of its supporters, although he denied their validity. These passages are underlined by Mendel in his copy of the *Bastarderzeugung im Pflanzenreich*. Moreover, the final section of Mendel's "Versuche" entitled, "Concluding Remarks", contains no less than a dozen references to Gaertner and three to Koelreuter, but none to Darwin. This section, as Mendel himself stated, is "a comparison of the observations made regarding *Pisum* with the results arrived at in their investigations by the two authorities in this branch of knowledge, Koelreuter and Gaertner".

On Unger and Mendel

Franz Unger's "Botanical Letters" were first published in serial form in the weekly supplements to the daily newspaper, *Wiener Zeitung*, in the summer of 1851. They aroused a storm of abuse from the orthodox right wing church newspaper, *Wiener Kirchenzeitung*. From the time of Mendel's arrival in Vienna until long after his departure the editor of this paper, Dr. Sebastian Brunner, wrote leader articles fulminating against Unger the "theological botanist". They culminated in the editorial of 29 January, 1856, entitled, "Isis Priest and Philistine", the result of which was to put Unger's academic career in Vienna in jeopardy. To prevent Unger's resignation 400 students of the Medical Faculty signed a petition calling for an end to Unger's persecution which was delivered to Count Leo Thun, Minister of Education. He intervened on Unger's behalf and Brunner was compelled to cease his attacks and publish an apology.

It has been suggested that it was the contrast between Unger's support and Gaertner's and Koelreuter's denial of evolution that stimulated Mendel to carry out his experiments (see p. 112). As the *initial* stimulus, however, we must recognise Mendel's own discovery of the uniformity of F_1 hybrids (see p. 114) which he made the subject of the introduction to his "Versuche". A powerful additional stimulus was undoubtedly given him in Vienna, namely, the desire to settle a question of crucial importance to the evolutionist—the source of variation. Unger's excellent work on the relationship between the distribution of plants and the nature of the soil and other external variables, which had been the inspiration behind the transplant experiments of Kerner von Marilaun (see p. 97), had already gone a fair way towards disposing of external conditions as the cause. This left one other possible source—hybridisation.

Darwin and Mendel

Since L. H. Bailey's reference to Mendel's "Versuche" was obtained from Focke's book, *Die Pflanzenmischlinge*, 1881, it can readily be seen that all three discoverers of Mendel owed their good fortune to Focke.

There is one other reference to Mendel as a hybridist. It is found in J. G. Romanes' article on "Hybridism" in the ninth edition of the *Encyclopaedia Britannica*, 1881. Unfortunately it is merely a mention of Mendel's name in a list of hybridists whose work was of recent vintage. Apparently Romanes obtained it from the historical section of Focke's book. When preparing the article he had enlisted Darwin's assistance, in order not to omit any worthy hybridists of whose work he was ignorant. Darwin advised him to read Focke and he recommended the historical section, at which he had glanced. He sent him his own copy (it was on sale in November, 1880), and Romanes followed his advice. When he returned the book to Darwin the pages which contain the section on the Leguminoseae were still uncut, as they have remained to this day. Thus did the Focke citation of the "Versuche" narrowly escape the eyes of Romanes and Darwin just as had the earlier citation by Hermann Hoffmann in 1869. Darwin underlined some of the passages in his copy of this work but passed over the brief Mendel citation.

In a recent publication (*Naturwissenschaftliche Rundschau, 18* (1965) 201–202) .Dr. J. Sajner has drawn attention to a report of Mendel's "Versuche" lectures which appeared in the daily newspaper *Brünner Neuigkeiten* on the 9th February and 10th March, 1865. In addition to mentioning Mendel's discovery of constant numerical relationships the report refers to the lively participation of the audience which it regarded as evidence of the success of the lecture.

Index

The letter *b* after an entry signifies that the page number which follows refers to a list of suggestions for further reading.